CONVERSATIONS WITH THE SPHINX

CONVERSATIONS WITH THE SPHINX

Paradoxes in Physics

Étienne Klein

Translated by David Le Vay

SOUVENIR PRESS

First published in France under the title
Conversations avec le Sphinx by
Éditions Albin Michel, Paris

This translation first published 1996 by
Souvenir Press, 43 Great Russell Street,
London WC1B 3PA
and simultaneously in Canada

Published with a translation grant

ISBN 0 285 63305 8 (hardback)
ISBN 0 285 63337 6 (paperback)

Phototypeset by Intype, London
Printed in Great Britain by
Biddles Ltd, Guildford and King's Lynn

This book is dedicated to
Don Quijote,
Raymond Devos,
Paul Valéry,
Richard Feynman,
and all those who dare to face the real
in what is strange, and vice versa

Contents

Introduction

One is never obliged to write a book.
Henri Bergson

Developments in modern physics have shown the inadequacy of common sense in the construction of its theories and the understanding of certain experimental results. In contrast to the everyday world, it opposes the extraordinary to the ordinary, difficult concepts to classical archetypes, astonishing descriptions to cosy appearances. It seems to have taken a malicious delight in overturning whatever is commonplace or just too obvious in every one of our concepts. Elementary ideas that seemed to reflect pure logic are revealed to be no more than the stale mental residues of the experiences of daily life.

The coherence currently acknowledged in physics did not come about by itself, or overnight. It is a slow product of the polemical reasoning that builds physics on the basis of its own self-questioning. But too often, in science, paradox is synonymous with parasite, and this, if it is not pure injustice, is at any rate sheer shortsightedness. Because they are anything but monstrous excrescences that blight the landscape of the history of thought, paradoxes deserve to be rehabilitated.

We need to bear in mind that science relies on the open character of the enterprise that enables it to call in question its own structures of thinking. Paradoxes are an integral part of its approach, which means that they should not be regarded as simple accidents along the way, which could be routinely avoided by greater attention or care. Paradox is intrinsic to the fabric of science. It is paradox that, by liberating habits of thought, provides orthodox reasoning with its dialectic. What would science be without its bursts of singularity? What would become of progress without means of propulsion? What would a thought be, deprived of the uproar of thought? Far from being lumbering monsters to be either put down or put away, paradoxes are the fuel of scientific progress. Truth in science, or what passes for truth, is nearly always a paradox resolved or an error rectified.

'O sacred errors, slow mothers, blind and sweating with truth,' said Victor Hugo. 'The way of paradox is the way of truth,' said Oscar Wilde, more blithely. These are two further reasons for taking a close interest in what the paradoxical visitors to the pages of this book may have to say.

The book is divided into two parts, intended to be complementary. One, entitled 'The physics of paradoxes', deals with the question of paradoxes from a broadly epistemological angle. The other, entitled 'The paradoxes of physics', deals more with physics proper. In part one, the concept of paradox is described in all its aspects, and an attempt is made to analyse the part played by paradoxes in science, particularly in physics. Part two outlines verbally – i.e. without equations – seven paradoxes of physics: the problem of wave-particle duality, Langevin's paradox of the twins in special relativity, the paradox of the dark night in cosmology (Olbers), the paradox of Schrödinger's cat in quantum physics, the famous EPR paradox raised by Einstein against the current interpretation of quantum mechanics, the violation of parity in particle physics, and lastly the paradox of the arrow of time. Readers impatient to make the acquaintance of these examples of physical paradoxes may proceed without loss to part two. Each case is presented there in a self-contained format, with all the relevant facts, and can be taken in the order dictated by individual curiosity. All the same, as a precaution against the baleful side-effects of overdosing on paradoxes – for example a wild and feverish fondness for every freakish fact, the notorious *traumatophilia* – readers are advised to space out their investigations. In order to assist them, we have made sure that part two contains as many chapters as there are evenings in the week.

PART ONE

The Physics of Paradoxes

1

A Consideration of Physics

A river is a personality, with its rages and
loves, its strength, its god of chance, its
illnesses, its greed for adventure.
Jean Giono

Nowadays, science is comfortably settled, having taken
a few centuries to impose and cement its reputation.
It has become usual to describe it as regulated by an
arid amalgam of objectivity and chill rigour, and this
is particularly true of physics, supposedly the most
rigorous and canonical of the so-called natural
sciences. Since it soars so capably through the strato-
sphere of pure ideas, people are inclined to see it, if
not as eminently consensual, then at least as exempt
from passions and quarrels. There are many who think
that it tolerates only a single type of discourse and
offers only a single vision of the world. They do not
imagine that it might contain within it notorious boil-
ing points or lasting divisions. It appears to present
an exemplary veneer that combines conformism with
dogmatism and even boredom. And there are those
who are not ashamed to admit that the spectacle elicits
a gentle yawn.

This conception of physics tends to reduce it to the
status of a simple body of knowledge, no more than a
compilation of facts, theories and methods collected in

learned texts, while scientists are likewise seen simply as those who endeavour to add a few more pieces to this particular collection. Thus science is perceived as a continuous synthesising process that can only be enriched quantitatively, by sheer accumulation, and makes its progress at the regular pace of a gentle gradient. No surprise, no imagination, no drama, is ever to be anticipated here.

This impression of serenity is not without foundation. Out of the successes that plead in its favour, science has developed an infatuation with its own superiority. It stands head and shoulders above the other disciplines that aim to grasp reality. Through the technical innovations that it has made possible, it has proved the efficacy of its methods. In this field it has no rival. No one any longer doubts the fertility of its concepts or validity of its statements, so much so that its content is readily likened to a higher form of truth.

Having become both honourable and authoritarian, today's science plays a part analogous to those that once were performed by theology or philosophy. Every sensible person must bow before the facts it brings to light and the conclusions it draws from them. It is next door to being a universal arbitrator. Because it offers the image of an objective verdict, it make it possible to put an end to argument. It is invoked in order to clinch the case. To claim that a fact has been scientifically proven is enough to pre-empt dispute. Enlisting science allows it to be concluded without more ado that 'it's a scientific fact', which is sometimes exasperating but always very practical.

Compared to the rigour of the sciences, philosophy is thought of as an archaic kind of knowledge, and its discourse as confused. When it is not accused of being a mental uproar marred by subjectivity, it finds itself shelved in the category of a priori systems. This destitution is not a new condition. Nietzsche wrote in *Beyond Good and Evil*: 'Science is flourishing today, good conscience is written on its face, whereas the decline into which all the new philosophy had gradually fallen, today's remnant of philosophy, incurs distrust and ill humour, if not mockery and pity.' Other writers contend that philosophy lost its sceptre and dignity on the day when the scientists rejected its guardianship. Paul Valéry remarked in *Variété V* that 'all unquantified knowledge has found itself struck by a pejorative verdict'. This effacement of philosophy has been progressive and is now complete. The progress of science has contributed so greatly to ridding philosophy of vain and false ideas that today an intellectual view, no matter how well argued, can be turned into objective knowledge only after trial by science, and not before. As the compulsory staging post for the validation of knowledge, the laboratory now functions as a filter for ideas. It alone is counted fit to tell the truth.

Yet science, like all living things, has its crises and its passions. Surely it is a human construct rather than a ruthless destiny. Kept irrigated, like a living being, by veins full of life, it reflects currents that in turn well up, do battle, gain the upper hand, persist, regress and then erupt again.

Far from resembling the changeless receptacle of our certainties, science constantly reinvents the twofold constraint that provokes and fertilises it: the heritage of its tradition, the world that it questions. Torn between these two poles, the history of science offers anything but the prospect of some long quiet river whose course is punctuated only by a few famous theoretical bends. Ideas, palpitations and convulsions dog it all the way, on a scale much grander than the narrative reconstructions are willing to suggest.

Without paradoxes, science would go round and round in circles. If it happens to be unable to flow back upon itself, that is because it is liable to so many chance encounters that complicate and muddle its path. Properly regarded, science exhibits a past that bristles with offshoots, zigzags, blind alleys, false hopes, and new paths that arise, supplant, transform. It springs from a bubbling, eruptive alchemy, which demonstrates that it is neither a stable locale, nor a perpetually orderly whole, nor a pool of stagnant ideas. Not all of us see things with the same eyes, and that is why the history of science occasionally runs riot. If reality came to meet us head-on, if we could enter into direct communication with things, mutual complicities, then science would be a simple read-out of the world, straightforward and spontaneous. But that is plainly not the case: science is no instant gift. It requires mediation, slow and fragmentary, because observing is not understanding, looking is not seeing, seeing is not knowing. Furthermore, not everything allows itself to be seen. The world around us is crammed with the invisible. Neither atoms nor quarks nor black holes can be seen with the naked eye. In order to think these objects it has been necessary to conquer the opaque immediacy of the world.

Far from dressing itself up in the clarity of the obvious, the universe conceals its laws, its ways and its machinery inside broad swathes of darkness. No glasnost here. Whether out of modesty, malice or playfulness, reality has no exhibitionist leanings. In order to expose it, the hunter has to undo or at least lift up the veil woven by a mysterious enchanter, keep looking for false trails, keep revising our ways of imagining what stands hidden in the mists behind the veil.

The real can create illusions, and that does not simplify the business of those who aspire to describe it. Within itself, mirages are always possible. Since obviousness is no guarantee of truth, it is possible to believe in fictions for a long time. Two examples concern our conception of the way the eye works and our version of cosmology. From Euclid until the end of the first millennium AD, vision was explained in terms of a 'visual ray', a beam emitted by the eye. For more than fourteen centuries this was the view of physics on the physics of the view. Not till the tenth century did an Arab optician, Alhazen, replace this notion by that of luminous rays given off by the object itself and coming to strike the eye, and his idea seemed quite strange. Did no look emerge from the smiling eyes of a child? They only received, and did not transmit? In fact the purely receptive nature of our visual organ had great difficulty in being accepted by the popular mind, and may still be unassimilated – we continue to speak of a 'piercing gaze' or 'looking daggers'. Reversing the real process, we are a lot more inclined to think of visual rays coming from the eye than of light-rays striking the eye from outside. Whereas the sole function of the eye is to keep on being showered by a rain of photons (drops of light), and the shining pupils of children are actually only transparent structures that admit these photons to the retina. So the gaze is only a charming illusion, which does not work the way we may imagine when we meet another's eye.

As for cosmology, it would be hard labour to make an exhaustive list of all the models of the universe that have been invented since the world began, and humans began to think about that fact. What is sure is that the modern version of cosmology has retained very few features of its ancient schemata. The recent theory of the Big Bang, which describes our universe as the outcome of a huge cosmic explosion, has preserved none of the explanations of Ptolemy or his predecessors. How many centuries and how much agony did it take to overthrow mythologies teeming with divinities, spirits and powers? How many bold ideas and slow maturings, to unravel the deceptive appearances of the sky? And for every right idea, how many wrong ones?

Our views about matter are neither spontaneously nor definitively correct, which means that matter is a matter of paradoxes. Whenever one of these crops up, science finds itself steering into difficult waters. Contradictions, doubts and passions come to the surface and stir up waves and surf. Speaking of waves, some readers may recall the episode of 'water memory' that in the autumn of 1988 set off all sorts of confused and impassioned reactions, and gave the media a field day. Some commentators avoided sensationalism and modestly attempted to remain 'scientific', but others resorted to gut reactions, whether for or against. In the end, the debate reached near-theological levels. It is by analysing situations like this that we come to understand the channels through which a fever can suddenly ravage the body of science. They contradict the image of a harmonious, tranquil science, often presented as a purely rational process, autonomous to the core. By what mechanism can reason, which is the the active heart of science, cease to speak with a single voice. What is the source of these incendiary arrows with the power to set science on fire?

If we invoke the goddess of Reason so insistently in the realm of science, and are pleased to make Reason its chief ruler, isn't it because we forget that nature has the power to surprise us? Not knowing everything about her, we describe as surprising anything that seems to us 'unnatural'. In this way we manage to give a meaning to the nevertheless contradictory concept of 'natural queerness'. As an alliance of two words of incompatible meaning, an expression like this is the opposite of a pleonasm – to be precise, it is an *oxymoron*. This apparently pedantic but etymologically humorous term comes from the Greek words *oxys* (sharp, keen) and *moros* (stupid, mad). A few examples are *gentle violence, eloquent silence, an open secret, encyclopedic ignorance.* Instead of cancelling each other, the two contradictory terms of an oxymoron blend to create the perfume of a mysterious poetry. On this point, we cannot sign off without citing Anatole France, who wrote in *Thaïs*: 'In those days the desert was peopled with anchorites.' Because it is the scaled-down model of antinomy, the miniature prototype of paradox, we owe it to ourselves, before concluding this excursion, to salute the magical, emblematic figure of the oxymoron.

Because there is a gap between things and the images we build of them, we have to be cautious in our use of the adjectives *rational* and *irrational*. The word *reason* has become as magical as a quality label, investing everything it is attached to with an aura of excellence. Yet the idea that reason is absolute and immutable is merely a doctrine, and an obsolete doctrine. What now seems to go without saying, because we have grown used to it, was not necessarily plausible from the start. In reality, nothing ever goes without saying. There is no immediate match between the rational and the real. What seems to us rational today once had to fight for recognition: it was not instantly recognised as such. Truth is not always truthful-seeming.

Rationality is a construct. It has the rhythm of a process. Any number of phenomena or ideas that in the past were considered untenable, illogical or paradoxical became rational and 'normal' once it was possible to integrate them into a coherent description. Hence our surprise when we observe that there is a certain fluidity among basic truths. A conceptual advance reshapes them, progress displaces them, a revolution demotes them to the rank of secondary truths.

Consider, from this point of view, the numbers that mathematicians call 'irrational'. What is so irrational about them? They were so named because they express the incommensurability, say, of the diagonal in relation to the sides of a square – an incommensurability that Greek geometricians perceived as scandalous. When they wanted to measure the diagonal of a square by means of the same units that measured the sides, they found to their dismay that it was impossible to find a unit of measurement that fitted both lengths. If the side was expressed as a whole number of times the basic unit, then the diagonal became 'irrational', and vice versa. Its length was given as the ratio of two whole numbers, in other words a fraction, whose denominator they could show to be both odd and even, which was clearly unthinkable. It was concerning the invention of these numbers, nowadays perfectly integrated into mathematics, that Hegel wrote: 'In its course, geometry comes up against incommensurable and irrational data where, if it wishes to go further in the act of resolution, it is pushed beyond the principle proper to understanding. Here too, as elsewhere, the terminology exhibits the inversion that consists in the fact that what is termed rational is what derives from understanding, whereas what is termed irrational is much more a start and a trace of rationality.'

In short, reason may lead to conclusions that seem at first irrational. Even when it sticks to performing the rigorous operations that it has itself defined, it may find itself walking up blind alleys or engendering contradictions. And as long as the latter are not solved, i.e. reappropriated and restated by reason, they are perceived as irrational.

Rationality does not know itself until it has stood the test of irrationality. It emerges from a dialogue with what lies beyond its own boundaries. Therefore its realm never ends where people think, because the irrational can breed in the interplay of reason, and when it is born, turn out to be rational. As is shown in the example cited by Hegel, rationality can, paradoxically, be built upon the irrationalities that it was first responsible for producing. So reason should be seen as the seething interior of a mobile sphere with an open periphery. Not knowing its ultimate frontiers, it never does anything but begin.

We have no means to distinguish a priori between what is rationally conceivable and what is not. If we had those means, the history of the sciences would have an aspect quite different from the one we know – perhaps it would simply not exist.

This obliges us to offer a brief critique of three current definitions of rationality:

1 The first states that *the rational is what seems normal*. Ever since Copernicus and Galileo, we have known that this amalgam doesn't work. It would only be valid for a superior intelligence freed from our handicaps and limitations. But for us who lack such 'vision', nature does not always stand to reason. The concept of normality evolves at the pace of our views about it. Thus it is not a constant; it fluctuates in time and space. The normal and the rational are relative.

2 The second states that *what is rational is what is effective*. In this case it must be admitted that quantum physics and relativity, which are both effective and astounding, are altogether rational. As both these theories had late and painful births, this means that reason has not always introduced itself to us in its rational guise!

3 The third state that *the rational is what tallies with the way that the world works*. This is a definition a posteriori. As the truth is rarely cut and dried, science cannot refer to a preconceived notion of the very rationality that it is its ultimate aim to define. Truth cannot be taken for granted as a prior category of thought. Man has to struggle very hard to become a rational animal. What he acknowledges as rational is not determined by any fixed external norm, but instead amounts to an issue constantly debated. Reasonable is changeable by nature and uncompleted by vocation.

So rather than routinely associating science with an idea of reason that it is hard to define, it is better to recognise that in its practice it is much more akin to a method. True, the ideal of rationality inspires its methods and acts as a fail-safe device, but that is not enough to move it forward. Intuition, convictions, imagination, flashes of genius – these are more effective driving forces. Evidence of the wealth and variety of human thought, they permeate the body of science with all sorts of wants, rivalries, scandals and prejudices that are closely bound up with its particular dynamism. Since science, like any other field, has its own received or preconceived ideas, then paradoxes become possible, and it is often due to these unexpected gatecrashers that brawling breaks out at the party of ideas.

But just what is a paradox?

2

What is a Paradox?

The dictionary dislikes gush.
Jacqueline Kelen, 'Praise of Tears'

The idea of paradox is neither plain nor simple. Whereas the traditionalist approach reduces it to contradiction, some have chosen to define it as 'truth standing on its head to draw attention'. The fact is that, far from denoting only antinomy, the concept of paradox is many-sided. The word is imbued with all the shades of meaning of its two Greek roots, *para* and *doxa*. *Para* is a prefix that signifies the adjacent, the out-of-phase, difference, singularity. It suggests the idea of distance in relation to something. When this distance is small, *para* conveys merely the idea of proximity, as for example in the adjective 'paramedical', which applies to everything on the boundary of medicine. But as this distance grows, *para* runs through the scale of notions that stretches from 'almost like' to 'quite the contrary', from a slight shade of difference to utter contradiction. Clearly the concept of paradox will be elastic and polymorphous. As for the word *doxa*, in any given culture it signifies the set of the opinions accepted without argument as natural, self-evident. The *doxa* is therefore prejudice and its periphery, it is the idea as overlord, everything that stands exempt from judgement. Within it can be recognised public and general opinion, the attitude of the majority, the voice of the natural, instant thinking. In sum, the paradoxical is whatever removes itself more or less far away from the power of the *doxa*. In the light of this etymological commentary, we can look at the three definitions of the word 'paradox' given in the *Grand Robert* dictionary:

1 Opinion, argument or proposition that conflicts with the commonly accepted opinion or likelihood.

2 An extraordinary, incomprehensible being, thing or fact that goes against reason, common sense and logic.

3 In logic, said of a proposition that can be shown to be both true and false.

These definitions highlight two great classes of paradoxes: those that run counter to common sense, and those that express a flagrant contradiction of logic (elementary or otherwise).

The former express the extraordinary character of a fact or conclusion. In the context of the natural sciences they are certainly the most interesting, since they may compel the revision of some earlier ideas. In physics they have played a key role by demonstrating the inadequacy of our familiar concepts. We shall try to define them more exactly later on.

Paradoxes of the second type are unacceptable from the viewpoint of logic, one of whose explicit aims is to unmask them. Once resolved, paradoxes are invaluable teaching tools that the philosophic schools of antiquity were skilled in using to perfect the art of reasoning and to avoid the pitfalls of rhetoric.

Logic (the *logos*) appears to be cast in a quiet role. It is often described as the orderly collection of fixed and inert ideas. Yet logical paradoxes really do exist: they clang like so many betrayals or profanations of the canon law of thinking. In the strict sense of the term, the logical paradox refers to a statement about which it cannot be shown whether it is true or false. The best-known (and without doubt the oldest) example is the liar paradox, traditionally attributed to the Greek philosopher Eubulides, of the Megarian school, in the sixth century BC. This paradox was originally expressed in the following manner. A liar is asked whether he is lying when he claims to be telling lies. If he answers: 'Yes, I'm lying', then he is evidently not lying, for if a liar states that he lies, he is telling the truth. Conversely, if he answers: 'No, I'm not lying', then it is true that he is lying, and therefore he is telling lies.[1]

The Greek logicians were intensely intrigued by the fact that a seemingly anodyne proposition could be so self-contradictory. Believe it true, and at once one must step back and believe it false. But no sooner does one decide that it is false, than the same step back leads straight to the idea that it must be true! This endless game of pingpong between true and false appears unstoppable. Could language itself be booby-trapped? Might its coherence be a hopeless quest? Many thinkers of Antiquity, such as Aulus Gellius, Chrysippus,

---

[1] Another and better-known version of the paradox has a Cretan called Epimenides assert that: 'All Cretans are liars.' This statement is logically contradictory if we assume that liars always lie, and those who tell the truth always tell it. Hence the sentence 'All Cretans are liars' cannot be viewed as a true proposition, because it would make Epimenides a liar and what he says a lie ... But if it is false, then Epimenides is lying by saying it. Here too the truth is lost in an endless closed loop.

Seneca and even Cicero, took a close interest in this fundamental paradox, which raises the problem known as 'self-referentiality': every language, every formal system, when it says something about itself, creates a structure comparable to mirrors reflecting each other into infinity. There is interference between the message and its medium, that is between the sentence and its own meaning. This phenomenon is very disturbing, because it means that simple ideas, utterly 'clear and distinct' ideas, as Descartes would say, can lead to logical dead ends, creating deep unease among those committed to pure thought. It was precisely because they symbolise contradiction that self-referential slogans saw their finest hour in the revolutionary 'May events' in Paris in 1968, on the walls of the Latin Quarter. 'It is forbidden to forbid'; 'Racists are inferior beings'; 'Stateless persons, go home!'

In his treatise *On Philosophy*, Aristotle looked into this question of self-reference, and his analyses prevailed for more than a thousand years. The story goes that Philetas of Cos, another Greek logician, lost his appetite because of this paradox, and even died too soon in the despair of being unable to resolve it. Never mind, though: this example remains an exception. In general, paradoxes are rather less dangerous to health than ordinary ideas . . .

One last example of paradox linked to self-reference, and even more traumatic than its predecessors, was Rüssel's paradox, stated early this century. We start with two simple a priori notions, that of the *whole* and that of the *part of a whole* – all things that seem very intuitive. Rüssel proposes to class wholes in two groups: 'normal' and 'non-normal'. He calls 'normal' any whole that does not contain itself as a part. For instance, the whole of physicists is a normal whole because, not being itself a physicist, it is not a part of itself (this does not mean that physicists themselves are all normal). Obviously a whole will be styled 'non-normal' if it contains itself as a part. Let us call the whole of all normal wholes N. The awkward question asked by Rüssel is: Is N a normal or a non-normal whole? Let us tackle both cases head-on. If N is normal it belongs to N, since N is the whole of all normal wholes; N is therefore part of itself, and hence – non-normal. Reciprocally, if N is non-normal it contains itself as a part (by definition), and is therefore normal because N is the whole of all normal wholes! Conclusion: *N is normal if, and only if, N is non-normal.* This seems absurd: starting from utterly simple ideas we have constructed a proposition that is both true and false. Our deepest convictions have suffered an outrageous assault. Even if it seems difficult, we are going to have to get used to such doctrinal upheavals. To the great displeasure of Descartes, the history of thought has shown no great indulgence to the doctrine of clear and distinct ideas. The fact is that simple ideas also have their no-go areas.

Unlikely though it seems, all this is not a game. In fact it was questions akin to those posed by the liar paradox that served as the starting-point in the work of the logician Kurt Gödel. They led him in the 1930s to formulate his 'incompleteness' or 'undecidability' theorem. This theorem, which convulsed the world of mathematics, is vitally important. It establishes that any formalised system based on arithmetic contains at least one proposition that is not decidable, which renders the whole system undecidable. More precisely, it shows that it is impossible to adopt a finite number of axioms such that every question is decidable. If the meaning of this theorem is generalised, it turns out that what used to seem the infallible tool of the human mind, deductive logic, really contains an impenetrable inner limit. Formulation cannot be self-enclosed within itself. In this connection there was talk about a 'foundations crisis' in mathematics. Actually there is no need to panic: Gödel's theorem has few practical consequences. All the same, it does reveal that mathematical reasoning is less perfect that was still believed at the start of this century, since the axiomatic method runs into a fundamental limit. Mathematics too cohabits with the arbitrary.

Logical paradoxes are also to be found at the root of what Kant called antinomies – questions that cannot be answered because two different lines of reasoning lead to opposite responses. As such, they embody the peak of dialectics (which is certainly very high). For example, the question 'Is space finite or infinite?' is an antinomy because it can be answered by saying:

– either that space cannot be finite, because we are incapable of conceiving that there is an end to space: whatever point in space is reached, we can always imagine going past it;
- or that space cannot be infinite, because space is a thing we can imagine (or else the word *space* would not have been invented), and we are incapable of imagining an infinite space.

Braced by this contradiction, Kant concluded that there is no rational answer available to the question of deciding whether space is finite or infinite. Even if this has few implications other than philosophical ones, intellectually speaking we must confess that it is still very vexing.

And what is to be said about paradoxes that express a simple flaw in the hypotheses or steps in a line of reasoning? If your small boy tells you how very glad he is that he doesn't like spinach, because if he did like it he would eat loads and loads of it, which would be unbearable because he can't stand the stuff, you would be aware of some incoherence in the air, but could you find the flaw? And what would you say to your bachelor friend who swears that he will only ever marry a woman who is smart enough to want nothing to do with him?

As well as the Kantian antinomy that we have just cited, there are plenty of other paradoxes that touch upon the concept of the infinite in mathematics. This is the case with the paradoxes of the phlegmatic Zeno of Elea, who lived in the fifth century BC. He is said to have been influential enough for Aristotle to acknowledge him as the father of dialectics, which is no small claim. A pupil of Parmenides, he shared with him the belief that the universe was a solid, uniform, motionless sphere, not subject to change, and to support this thesis Zeno set about discovering a number of paradoxes concerning motion. He tried in particular to show that the idea of motion along a continuous line led to absurdities. Thus he taught that in order to travel a given distance it is first necessary to cover a half, then a quarter, then an eighth, a sixteenth, and so on. After each move there remains a distance to be covered, which never reaches zero. Zeno concluded that it is impossible to cover a given distance, and hence that motion does not exist. This conclusion, frustrating for runners, is called the paradox of dichotomy.

As all of us feel in our bones the existence of motion, we have trouble accepting it. It is also related that when the Eleatics defended Zeno's theory in public, and concluded as he did that motion is merely an illusion, Diogenes the Cynic displayed his scepticism by getting up and going for a stroll.

Zeno's paradox results from the (erroneous) conviction that it must take an infinite time to cover an infinite number of stages. In Zeno's era the notion of convergence, so essential here, and more precisely that of an infinite convergent series, was unknown – and with good reason, as it took nearly two thousand years to formulate an acceptable solution to the paradox. It led to the heart of the concepts of space, time and motion, which proved in the process to be far more subtle than they seem. But let marathon runners and other road users take comfort. Even if one thinks of these modern Sisyphuses as winning useless victories, their efforts are not in vain. When they run, not only do they exercise their muscles, they also advance!

All paradoxes are not equal. Not all of them have the same opacity or depth, the most abstruse are not always the most fruitful, and the easiest to expound – the liar paradox, or Zeno's – may be the hardest to resolve. A paradox is rarely soluble in the light of its own expression. For example, those that are referred to, very disparagingly, as sophisms are not on that account futile, even if a little rigour will usually suffice to resolve them. They sometimes compel a complete reconstruction of the systems in which they develop.

Some paradoxes are both very primary and very spectacular. The conclusions they force us to accept are true, but contrary to the surest intuitions. For example, imagine a ribbon stretched all the way round the earth, say at the Equator, and pulled tight, as if to wrap a present. Then suppose that we add to this taut ribbon a metre of slack, which has the effect of loosening it slightly. Now consider the following question: knowing that the radius of our planet is 6,400 kilometres, how far would that ribbon have to be raised above ground-level, using that metre of slack, to make it taut again? Usually the figures suggested on the spur of the moment are ridiculously low: a millimetre here, a micron there. The right answer is $\frac{1}{2}\pi$ metres, or about 16 centimetres. Better still, that value does not depend on the size of the object round which the ribbon is stretched. The result is the same whether we perform the experiment around the earth, or a pebble, or even round an atom. If you tell this to your friends, some will refuse to believe you and will press you to provide a demonstration (mathematical but elementary) of what you say. The result is so different from the conjecture of intuition, the still, small, authoritarian voice that acts as our spontaneous guide, that it can only be swallowed after much argumentative ado. This is how paradox becomes the most clinching and effective means of conveying the truth to the distracted.

All logical paradoxes are fascinating. Because a stream of books has already been devoted to them, we shall not raise them from here on, but this abstention will do us no harm, for the paradoxes of physics are no less absorbing than their logical counterparts.

The paradoxes of physics most often express the extraordinary character of a fact, a conclusion or a discovery. But this definition is too general, and possibly incomplete. We shall start by distinguishing three types of physical paradoxes:

1 *Those that throw light on a discordance between a theory and a particular experiment, or else between several concurrent theories.*

These paradoxes are a matter for experts. They carry the heaviest implications, because their resolution sometimes imposes far-reaching changes on the theory that bred them. It is certainly these that Joliot-Curie had in mind when he said: 'The further an experiment is distanced from a theory, the closer to the Nobel Prize.' These paradoxes do not refer directly to the blind-spots of common sense, because they involve theories that may themselves conflict with the standard concepts.

2 *Those that convey a surprising fact that offends common sense.*

Even when perfectly well understood by science, and integrated into it, many of its findings defy common sense. It is no accident that the Greeks used two different words to draw an early distinction between the discourse of opinion (*doxa*) and of science (*episteme*). In reality, almost all the findings of science offend common sense. In this respect they amount to paradoxes in the first sense of the word, even if they are quite often only an attenuated form of the paradoxes of the previous type. They count as paradoxes for the general public, but no longer for the experts.

3 *Those that result from a contradiction or internal inconsistency in theory.*

These come closer to the logical or mathematical paradoxes, but most of all they ask a startling question (with loud Platonic echoes) about the relationship between mathematics and reality. How is it that mathematical logic is the proper tool for gauging the internal consistency of physical theories?

By nature polymorphous (but seldom perverse), physical paradoxes contain and blend these three components in varying proportions. The abundance and diversity of their appearances give an idea of the influence they have had on the development of scientific discourse and thought.

3

Paradox, or the Catalysis of Thought

> Two ideas are always needed: one to kill the
> other.
> *Georges Braque*

Science used to be Aristotelian, and now it is not. It
used to be Newtonian, and now it isn't quite. Science
has mutated several times, in irreversible fashion, and
if we seek the origin of these mutations we find one
or several paradoxes which have acted as more or less
powerful detonators. Even if these paradoxes are not
explicitly referred to at the time when crises develop,
even if they are relegated at first to the less explosive
rank of *anomalies*, they are at least 'in the air'. They
can always be identified after the event, and be perceived as making necessary the mutation with which
they were associated. So we demonstrate the way in
which science has been constructed and get rid of
the contingencies that preside at the birth of ideas.
Paradoxes are abundantly capable of delighting all
those who are interested in the life and history of
ideas.

It is through the play of paradoxes that what has
been believed to be true may cease to be altogether
true. Without them, established theory would be

absolute, static, definitive; no case would ever be re-opened, there would be no possibility of considering new ideas, and therefore no progress. The life of ideas would be bleak, with no challenges, and hence no flying sparks.

The human mind is conservative. It grows accustomed to the ideas it deals in, constructs prejudices that it takes for principles, and always ends by loving what it believes. As Bergson rightly said: 'Our mind has an irresistible tendency to consider that the clearest idea is the one it most often employs... With usage, ideas acquire undue value' (*La Pensée et le mouvant*). It changes only when compelled to do so, and paradoxes are precisely the compulsions that are capable of disrupting cerebral determinism and preventing stagnation of the intellect. So they have both a disheartening side, because no one likes to be forcibly disturbed, but also a fascinating side, because what is not understood is attractive.

Francis Bacon compared the human mind to a warped and dirty mirror whose surface needed to be cleaned and polished before it could produce true images of nature. The paradoxes of physics are reflections of these irregularities; they embody in polemic fashion the Platonic duality of the model and its image. Appearing at the intersections of the forms of thought and the structure of things, they reveal the imperfections of the human mind whenever it confronts reality. But, at the same time, they motivate it. As vectors of the creative force of doubt, paradoxes are puzzles that tease and rack the brain, provoke the neurones and stimulate the synapses. By indicating the places where reason is abused, they put it to the test; by upsetting its certainties, they sharpen its powers; by stimulating its continual uprooting, they carry the seeds of all its victories. For all these reasons, paradoxes deserve to be the darlings of the mind.

Only critical situations are propulsive. A victorious idea always ends with nothing to confront. Doomed to repeat itself to infinity, it finds only what it has given to itself. Hence, the only real intellectual function can only be the kind that comes to grips with contradiction, with contraries, with fault-lines, and which always runs counter to the obvious. The way the human mind works is certainly a dialectical: without contradiction it makes no progress. It only shakes off its inertia through a counter-dynamic. The true objects of thought are therefore objections. In particular, if science advances, it is because both experiment and theory run into obstacles that prevent them from stagnating. Paradoxes are the enemies of static regimes of the mind.

Only an idea in a state of alert is capable of evolution. To understand, it is first necessary to understand that there is something to be understood. Science would not advance without something to spur it. Within it, progress and contradiction are twinned as the poles of a dialectic articulated around paradoxes, which act as catalysts. Paradoxes foreshadow revolutionary developments in science. Whenever, in whatever discipline, a problem appears that cannot be resolved within the conceptual framework in which it originated, the mind experiences a shock. It is this shock that may compel it to reject the old structure and to make a new one. Most scientific ideas are due to this process of intellectual mutation. As we have seen, the concept of a convergent infinite series arose from Zeno's paradox. In mathematical logic, the seeds of Gödel's theorem were antinomies like that of the liar paradox. In physics the paradoxical outcome of the Michelson-Morley experiment shattered the 'ether' concept on whose basis the theory of electromagnetism had been constructed. The dual concept of light as particle and wave forced physicists to re-examine the concept of determinism and led them to quantum mechanics. The paradox of Maxwell's demon gave rise to the profound intuition that the apparently disparate concepts of information and entropy are in fact intimately linked. And so on. The list goes on and on.

Paradoxes have done great service, for it is only through its ability to stumble over itself that science has become science. The least of rewards would be to think of inscribing over the entrance to the temple of ideas: 'To the great paradoxes: from Thought, in gratitude'.

4

Paradox and Contingency

There is an atmosphere of ideas.
Honoré de Balzac

Paradoxes are historically bound to their own epoch and context. They obey the absolute laws of relativity. The same fact does not convey the same reality and does not have the same meaning if it is perceived by a Cro-Magnon man, a Greek philosopher of the sixth century BC, a medieval scholar or a modern physicist. In consequence, paradoxes can be dated. To some extent, it is through them that knowledge has a history. They are associated with the turning-points that 'historicise' physics. There is the physics of before such a paradox, and the physics of afterwards. To understand the history of physics is therefore to make an inventory of all the paradoxes it has had to tackle, to recall the headaches they inflicted on physicists, and to understand the solutions they proposed. All this is deeply instructive. By analysing paradoxes in their context, one becomes aware that present-day physics is just the tip of a very long history of errors overcome, and it is clear that it is all these leapfrog advances that constitute its strength.

Paradoxes *drive* physics. They bring to life the ideas its practice develops, and also make them mortal.

Søren Kierkegaard knew what he was saying when he wrote:

> One should not think ill of paradox, the passion of thought. A thinker without paradox is like a lover without passion, a handsome mediocrity. But the essence of every passion taken to its extreme is always to will its own ruin. Similarly, the supreme passion of reason is to seek out an obstacle, even if this causes its destruction in one way or another.

This message seems especially true of basic research. The basic physicists of today dream of a good paradox that would light the blue touch-paper and enthuse their researches, ready in the last analysis to change their entire systems of thought. On the other hand, it seems that applied physics (and more generally, technology) can grow only at some distance from paradoxes, and from all the commotion they arouse. To have hopes of mastering a well-defined field of applications, it is better to have stable, reliable models, structures to be trusted.

The fact remains that the cathartic effect of paradox mentioned by Kierkegaard tends to temper the arrogance of all forms of discussion, including the scientific. Paradoxes prevent science from setting up as a dogma. By the regularity of their appearances, they make it foreseeable that tomorrow's understanding will come partly by way of negating what is said today. When necessary, science can oblige itself to offend its own pieties.

The clarification of science by paradoxes is not a one-way process. By a fair exchange, the history of the sciences brings out all the aspects of the concept of paradox. So the clarification is mutual. The paradoxes of nature reveal the nature of paradoxes, and this mirror effect makes them doubly absorbing.

As the inconsistency of a thesis depends mainly on who is listening to it, paradoxes naturally have a subjective dimension. Each mind has its own telescope, each brain has norms that mould the thoughts and feelings it produces. Even within the community of physicists, apparently so homogeneous, some see a paradox where others notice nothing out of the ordinary. We have only to quote the case of the founding fathers of quantum mechanics – men like Planck, Einstein, de Broglie, Schrödinger, Bohr, Heisenberg, Dirac – who clashed about what meaning to assign to this new theory. Their philosophical assumptions, and even their emotional makeup, directly influenced the interpretations that they put forward or accepted. Really there are no paradoxes without prejudices, just as etymology proclaims.

Similarly, it is utterly commonplace to say that the recognition or non-recognition of a paradox is dependent on the community to which it is presented. Not everyone lives in the same world. Thus, for example, one knows that there exist types of reasoning seen as classical in India, for which no equivalent can be found in our treatises of logic. At the start of this century, a great Indian mathematician, Srinivasa Ramanujan, arrived for a prolonged stay in Europe. The European mathematicians who spent time with him were struck when they found that the style of his arguments was very different from their own. That is because neither methods nor principles are everywhere the same. Inuit or Sherpas obviously do not have the same frame of reference as the inhabitants of the Bastille quarter, or the same bees in their bonnets. And we know that a simple mountain range may act as a buffer against truths, and prevent their spreading. 'Truth this side of the Pyrenees, error beyond,' said Blaise Pascal.

This state of affairs has its implications for the practice of science, as can be seen in the following example, taken from the history of cosmology. Until the astronomer Edwin Hubble demonstrated the recession of the galaxies in 1929, the idea that the universe is eternal and unchangeable was strongly rooted in Western thought. Even Einstein, not usually scared by new ideas, did not dare to question this dogma, although his own equations, those of general relativity, had been begging him to do so since 1915. Instead, he added to them a term, necessary in his view, whose sole function was to block artificially the expansion of the universe that his equations appeared to require. His background did not allow him to imagine a cosmos that expanded like a balloon. By doing so, he sidestepped what would certainly have been the most fantastic prediction in the history of physics, that of the expanding universe. We can always imagine that if Einstein had been Chinese, he would have had fewer scruples, since, unlike the Western tradition, Chinese tradition does not rule out an evolving universe. There is a well argued account of such a universe in a book entitled *Wu Li Lun* ('On Physics'), dating from the third century BC.

But it is not only geography that modifies our a priori ideas. Reference frames change, experience differs from one group, one profession, one culture, to another, as if they were separated by conceptual Himalayas. What is obvious to a physicist may seem inconceivable and absurd to the uninitiated, and vice versa. For the layman there are the 'ravings of the initiated', for the expert the blinkers of the layman. That is why, when moving from one circle to the other, one is obliged to explain what previously went without saying. Take the example of the roundness of the earth. In view of the fact that this was demonstrated long ago by Eratosthenes (around 250 BC), it might be thought that it is now generally taken for granted. In fact, the notion that the earth is a sphere still worries some people, and particularly children. As long as the difficult concept of the force of gravity remains unassimilated, how can one fail to think that 'if the earth were really round, the people on the other side would fall off'?

5

Common Sense and its Limitations

He followed his idea. It was a fixed idea, and
he was surprised to make no progress.
Jacques Prévert

Common sense designates the geometrical site of our
prejudices. It is the consensual element in our tempera-
ments, the lowest common denominator of our sensi-
bilities, 'the best shared thing in the world', as
Descartes said in his *Discourse on Method*. If it is
rightly found guilty of reducing the mind to a free-
wheel, and thought to its own inertia, that is because
it deprives them of the element of reflection that would
render them dynamic. Because it has the bad reputation of supplying ready-made replies, it is not surprising that it is suspect number one in any case of
paradox and guilty almost every time. By inhibiting
reflection, it conditions our reflexes. By prefabricating
our prejudices, it channels our reactions. By construct-
ing norms, it determines our manner of being surprised. From all these angles, common sense famously
interferes in every issue of paradox.

Common sense is a nondescript little imp. The tyranny it exerts over our judgement is sly, discreet and
anonymous. It regularly amuses itself by deceiving us.

True, our naivety has few excuses. Many philosophers, and not the lesser ones, have alerted us against its inadequacies, revealed its shallow nature, and condemned its tricks. This has gone on for some time now, and in every possible tone. Ever since Parmenides' poem *On Nature*, only fragments of which survive, opinion has been tried and thoroughly condemned, described as 'the antipole of the heart, without a tremor of Truth' (*Fragments*, 1, 29). As for the opinions of mortals, 'one cannot rely on any truth therein' (1, 30). Parmenides was one of the first to tell us that one should not believe too much in beliefs, opinion is not truth, and our senses are full of inexactitudes.

Immanuel Kant was no more indulgent towards common sense. He saw it as a desperate last resort that one will take good care to avoid as long as any critical sense remains. In his *Prolegomena to any future metaphysics which may present itself as science*, without mincing his words, he rebels against what he sees as the dictatorship of common sense, and rejects the false hierarchy it promotes:

> To appeal to common sense is to allow the most vapid gossip-monger to safely defy and oppose the sturdiest intelligence. More closely considered, this appeal amounts to relying on the judgement of the crowd: an approval fit to make the philosopher blush: a matter of triumph and pride for the popular buffoon.

Common sense is all the more insidious for being nearly always legal tender, to use a monetary term. No one completely escapes it, and of course, that is its definition.

Paradox maintains a changing and subtle relationship with truth. If it were simply an opinion contrary to an experimental truth, then it would always be false, but it is only a thesis contrary to common opinion; and since common opinion may be wrong, paradox may be right. Therefore it is not error, it is not the opposite of truth. Between them there is not opposition, but a symmetrical relationship. If there are paradoxes in our truths, there is also truth in our paradoxes.

Of course it would be silly to deny that common sense is of great practical use in our everyday life, or that its absence will hamper those who lack it, sometimes quite cruelly. It has a functional utility which is essential to us. What would the activity of thought amount to if we did not start with a small stock of prejudices to feed it? What would our brain do without grist for its mill? Yet we are bound to recognise that the sphere of validity of common sense is very limited. It is a quality of character rather than of mind. To be precise, it only becomes a quality of mind provided that a higher disquiet keeps watch on it, and if necessary reminds it of its infinite ignorance.

Common sense, even under strict supervision, can claim to lead to the truth only by a roundabout path. This case may be summed up by saying, at the conclusion of a curious syllogism, that common sense leads to the truth since the truth is born of paradox, and paradox is defined as against common sense. But all the way along this tortuous path, common sense must be worked upon, criticised and adapted, made aware of itself and its inadequacies. Otherwise, it will never be more than the sleepy routine of the lower levels of the mind.

Our familiar concepts only sum up the possibilities of common sense. Being subject to the same defects, they deserve the same criticisms. In particular they are often too archaic, or too simple. To recognise this is already to bring out their intrinsic limitations, since what is simple is only what has been simplified. Because of this, the tactic that consists of approaching a problem in simple terms is bound to come up against a limit. The simple always ends as the simplistic. We must act in such a way that 'everything is as simple as possible, but no simpler', Einstein advised. If the limit of the simple is transgressed without noticing, then you end with false results. Initially convenient and attractive, simplification soon becomes misleading and dangerous. Every attempt at popularisation runs into this limit of the simple: if the simple is false, and the complex inexpressible, how is science to be expressed without betraying it?

In the nineteenth century, the historian Ernest Renan was already saying that 'in science everything is fertile, except good sense'. The twentieth century has proved him a thousand times right. Since the year 1900, physicists have had to perform some extraordinary leaps of intuition to explain the new facts thrust upon them. In discovering that there are no more obvious data on the borders of the invisible, physics has had to shed its old reflexes and renounce many habits based on what had become common sense. The majority of its gains are victories, not *for* good sense, but *over* it. We have lost count of the situations where nature, the ultimate referee, has sent it back to the changing-room. In science, good sense attracts red cards.

Nietzsche said that everything decisive is only born *despite*. Each new truth originates *despite* the evidence, every new experience originates *despite* current experience. This is so true in science that it is fair to speak of an opposition between opinion and science. A famous pronouncement of Gaston Bachelard's asserted that 'science is absolutely opposed to opinion. If it should happen, in some particular instance, that it substantiates opinion, it is for other reasons than those on which the opinion is based, so much so that, in law, opinion is always wrong. Opinion thinks badly; it does not think at all; it translates needs into knowledge ... Nothing can be based on opinion: one must begin by destroying it. It is the first obstacle to be overcome.' (*The Training of the Scientific Mind*). For Bachelard, opinion is defined not only negatively, by its absence of information, knowledge and reflection. It is a positive obstacle to scientific knowledge. Bachelard was not afraid to define scientific experiment as 'an experiment that contradicts common sense'.

Let us take two of the simplest scientific findings: The first tells us that we have to accept the far-fetched idea that the earth is a ball rotating at a giddy speed on its axis and orbiting the sun, even though it seems to us perfectly motionless. The second explains that, even though all the movements we can observe on this earth slow down and stop spontaneously, we must agree to understand them on the basis of the principle of inertia, i.e. according to the ideal of a uniform motion which, if nothing exerts a braking or accelerating force, goes on for ever.

These conclusions are neither crude nor hasty. They are not the result of a simple reading of the world around us. Science had great difficulty in pinning them down. The principle of inertia which has just been referred to states that every body perseveres in the state of rest or of uniform motion in which it finds itself unless some force acts on it and compels it to change its state. This principle is not at all intuitive, except conceivably for virtuoso ice-skaters, because intuition tells us instead that movement is essentially linked to action. Without some force to offset its slowing down, it stops. That is how the Greeks, who had noted that in order to draw a cart at a constant speed, oxen must exert a constant force, concluded that all movements required a cause, and that the natural tendency of a body in motion was to return to a state of rest. This idea, according to which all activity implies that something is acting, oozes with good sense and corresponds to what we observe when forces of deceleration come into play. It was maintained in Europe for centuries without much discussion, since the great Aristotle himself had written in his *Mechanics*: 'Everything that is moved is moved by something. The body in motion stops when the force that pushes it no longer acts so as to push it.'

Elementary, my dear Aristotle! So elementary that it took centuries, and all the genius of Galileo, to counter this unanswerable argument of Aristotle's with the principle of inertia, which was to turn physics upside-down. Thanks to him, something that had never been observed (motion that never stopped) became the guiding principle to explain the observable (that movements do die away). On these grounds, it was one of the most important victories in the history of thought, and unquestionably marked the true beginnings of physics. Its importance was cardinal. For the first time it was demonstrated that one could not always rely on intuitive conclusions based on direct observation.

Galileo had understood that evidence can be the bane of science. Against all the evidence.

6

The Senses, Science and Ourselves

'But daddy, Einstein says . . .'
'Shut up about your Einstein, I don't want
any trouble with the neighbours.'
Fernand Raynaud

The world of the senses does not tell us the essence of the world. There is clearly a break between sensory knowledge and scientific knowledge. Our physiological senses deceive us, for they make us perceive and feel a world that is not the world of physics.

Whenever we leave the conditions of our usual environment, surprises spring up and cast doubt – sometimes quite violently – on what we had grown used to believing. Every attack of amazement should remind us that we are no longer entitled to identify scientific knowledge with human habits. These are a very sorry guide, avid for routine and convenience, a great fan of the familiar, not greatly tempted by the breaking of new paths. Modern science has exposed their appalling poverty of invention. To understand the world, one must forgo the comfort of obvious assumptions which are both facile and sterile.

The extrapolation of our habits of thought beyond their context does not yield good results, and physicists have had to invent more unusual explanations. One example is the special theory of relativity formulated by Einstein in 1905, which introduced the concept of 'space-time', replacing separate concepts of space and time. If you change the frame of reference in a space-time continuum, time is partly transformed into space and space partly into time! This idea is in no way part of our intuitive ideas, and is even foreign to all traditional philosophies. For Kant, for example, as for all the physicists of his era, space and time formed two entirely distinct concepts, so that he had no reservations about calling space 'the form of our external intuition' and time 'the form of our internal intuition'. These separate concepts were discarded by relativity when it raised space-time to the rank of a four-dimensional continuum – a concept that no longer contains anything familiar. And this is not to mention the general theory of relativity, which appears to account for gravitation. According to Einstein, its author, gravitation is not a force that exists itself between the various material contents of the universe. Rather it is a geometrical property of the universe itself. This insertion of gravitation into space-time forces it to distort and warp itself, in other words to become *curved*. It is within this curved space-time that space, time and *also* matter have tangled intricacies that no one before Einstein had suspected.

In particle physics, there are even more striking things (the word is well chosen: most of the experiments consist of organising particle collisions of extreme violence, and then seeing what particles emerge from the point of collision), such as the *phenomena of creation and annihilation* of particles. Imagine two particles, for example two protons, that we induce to collide. After they have interacted, the two protons separate, but their motions have partly materialised, since new particles have appeared in accordance with the equation $E = mc^2$. It is as if pure motion has been transformed into matter! This transformation of a property of an object into an object does not appear on the list of received ideas. We are much more inclined to believe that there are objects on the one hand, and the properties of these objects on the other, and that normally these are two kinds of things which are not transformed one into the other. This explanation of the creation of articles is staggering. It is rather as if someone told us that the speed of Carl Lewis could give birth to another sprinter (which would complicate the judging of races), or that the height of Annapurna might give birth, not to a mouse, but to a new mountain (which would complicate the work of cartographers). Weird is the world! And yet it is just this kind of thing that physicists observe when they work with particle colliders. In these, energy is effectively transformed into matter. Particles are created out of energy and then disappear by forming radiation or other particles.

These two examples show how our vision of the world has changed formidably in a few decades, and at a rate unprecedented in history. Have these upheavals overflowed the strictly scientific framework in which they arose? Does our era live in harmony with the knowledge it produces? Not really. True, science dominates, but the social, political and economic ideas that prevail today were nearly all shaped, consciously or not, by a vision of the world based on the findings of nineteenth-century science. We continue to see science more or less as our grandparents saw it. Take the example of determinism. We now know that it is no longer what it was in the last century. It waned at the sight of the atom, before decaying into Heisenberg's uncertainty principle. In his *Philosophical Essay on Probabilities*, dated 1814, Laplace believed it valid to write that 'an intelligence which, at a given moment, knew all the forces which animate nature, and the respective position of the creatures that compose it, if moreover it was vast enough to subject these data to analysis, would embrace in the same equation the movements of the largest bodies in the universe and those of the smallest atom; nothing would be uncertain for it, and the future, like the past, would be present to its gaze.' Today, no one any longer believes in this idea of a mathematician's possessing an equation that could link the past and future of all movements. Quantum mechanics in particular has shown the arrogance of such a claim. Yet we cheerfully continue to link science and determinism in the confusion of the same pleonasm. One day the clocks will have to be put forward.

There is a tendency nowadays to confuse science with technology, possibly because science has become much more a source of power than of knowledge. In particular, it is a pity that philosophy does not take sufficient interest in its advances, thus losing contact with an entire field of contemporary knowledge. This mutual indifference has led to a split. Many scientists, particular physicists, do not take enough interest in the philosophical debate that deals with the dual problem of the origins of science and the consequences it brings about, as if science did not deserve to be considered both with subtlety and in its totality. As for the philosophers, they pay too little heed to the upheavals in science which have occurred under their very eyes, as if an altered perspective on the nature of reality could have no repercussions in their discipline.

Yet many philosophical questions underpin the approach of physics, and conversely, physics has conceptual implications that cannot in honesty be ignored. Consider the exemplary case of quantum physics. From its first faltering steps at the start of the century it troubled the minds of its founding fathers, who soon realised that there was something disconcerting about it. Coinciding with the intellectual revolution that turned upside-down most branches of culture (painting, music, literature), it altered our description of the physical world. Like cubism, serial music or surrealism, it provided in its own field an answer to a severe crisis of representation, in which classical physics found itself incapable of describing the microscopic phenomena that were starting to become known. The founding fathers of quantum mechanics each in their own way tried to add their voices to the great debate that accompanied its birth, but it must be admitted that these initial discussions, which are still far from concluded, were gradually overshadowed by a resigned consensus about the efficacy of its methods.

Quantum mechanics differs from classical physics in fundamental aspects, and particularly in its poor absorption into familiar concepts. Classical mechanics had reached its maturity in the nineteenth century, drawing deeply on the sources of common sense that two centuries of Newtonian practice had forged and honed. Those were still the good old days when the universe still resembled a vast clock whose elementary parts were still elementary. Atoms were particles, light behaved like waves, the equations looked healthy. To explain the kinetic theory of gases, physicists could make do with talking in terms of billiard balls. In short, it was still thought possible to entertain a mechanistic explanation of the world. The concepts introduced by Galileo, Kepler and Newton had succeeded so well in explaining the motion of celestial and terrestrial bodies that it seemed that they could serve as a basis for the solution of every problem posed by physics. Notions drawn directly from our daily experience had been stood on intangible pedestals. This was the case with speed or acceleration, which are familiar to us all, even to those who cannot formulate them in mathematical terms. (It is unnecessary to master the notion of derivatives to understand that a Formula One racing car goes faster than an ox-cart.) In classical physics, the understanding of natural phenomena often coincides with current sensibility, which has the benefit of putting the intellect at ease with the images it canvasses. This comfort lasted only two and a half centuries – the time that elapsed between Newton's publication of his *Principia mathematica* in 1687 and the writing of Schrödinger's equation in 1925. The classical edifice literally exploded in contact with the atom. The repeated assaults of the *quantum* denounced it as simplistic, destroyed its foundations and damaged its convictions. The mechanistic account of the universe did not survive the axe-blows methodically aimed at it.*

Quantum physics is both limpid and enigmatic. It is limpid because its formalism is clear and its predictions

---

*There is an untranslatable pun here. 'Axe-blows' translates the French *coups de h* because *hache*, which means axe, is also the sound of the letter *h*, which happens to be the symbol for Planck's constant. (Tr. note.)

have never been experimentally refuted. It is enigmatic because no one understands it. For over sixty years it has not received an interpretation far-reaching enough to eliminate all the paradoxes that have been opposed to it. Since the 1930s there has certainly been a going 'orthodox' version of quantum mechanics, called the 'Copenhagen interpretation', but this interpretation has not led to unanimity among physicists. It was fiercely debated at the time of its invention during the autumn of 1927 at the Solvay conference in Brussels, particularly by Bohr, who defended it, and by Einstein, who criticised and even tried to squelch it. Einstein devised numerous thought experiments with the aim of detecting its possible inadequacies, but it held together, and seemed to conform to these experiments as far as could be seen.

The Copenhagen interpretation stemmed from a paradox: every physical experiment, whether it deals with the phenomena of everyday life or with atomic phenomena, is necessarily described in terms of classical physics. These form the language we use to describe the conditions in which we stage our experiments and communicate their results, and it is impossible to replace them with others. Yet the application of these concepts is limited by what are called Heisenberg's uncertainty relations. What then is to be done? The Copenhagen interpretation claims to respond to this dilemma, but it has not succeeded in convincing everyone. While the formalism, operant framework and mathematical apparatus of quantum theory are today universally accepted, debates still rage about its interpretation and philosophical implications.

Some predictions of quantum physics are surprising. 'Those who are not disturbed by discovering the quantum theory have not understood it,' remarked Niels Bohr, who in any case had adopted the slogan 'It's not crazy enough' to reject (with varying success) every idea that seemed to him too conservative. The fact is that the simple ideas, those that prove themselves every day, have been denounced as false or irrelevant on the microscopic scale. By prohibiting pictorial images, quantum mechanics removed from the relations between reality and knowledge the false impressions they conveyed at the beginnings of the scientific era. Because of it, the hope of building science with the raw elements of the external world has been disappointed. No more Meccano. Physics has turned away for good from the materialist line it thought it possible to follow in the nineteenth century. But one may well ask whether our culture has managed to incorporate all these New Deals.

Can we be sure of having really abandoned the idea that the universe is a great clockwork mechanism, like a watch, but on a larger scale?

7

The Thematic Mass of Mathematics*

Are words' roots square?
Eugène Ionesco

Ordinary language can no longer be used to speak about the atom, so much so that when one follows the efforts of the founding fathers of quantum mechanics to understand it, one wonders whether the fundamental role of the atom is not to force people to do mathematics! Should readers need convincing, they are invited to cast an eye over the page of calculations we reproduce in Figure 1, taken from a book on quantum field theory. For modern science there is no longer a material object as foundation, but structures, forms or mathematical symmetries, in other words purely intellectual creations, which alone seem capable of reflecting the cohesion of reality.

Mathematics obviously did not wait for the advent of quanta to insert itself into the formulations of physics and invade it with symbols. Ever since Galileo, these have been allied with reasoning, and almost at once their application introduced a dichotomy between

Another untranslatable sound pun: 'L'amas thématique des mathématiques (Tr. note.)

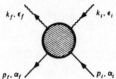

k_f, ϵ_f k_i, ϵ_i

p_f, α_f p_i, α_i **Figure 5-2** General kinematics of the Compton effect.

The connected S-matrix element (c stands for connected) is

$$S^c_{f \leftarrow i} = -e^2 \int \int d^4x \, d^4y \, e^{i(k_f \cdot x - k_i \cdot y)} \langle p_f | \, T : \bar{\psi}(x) \not{\epsilon}_f \psi(x) : \, : \bar{\psi}(y) \not{\epsilon}_i \psi(y) : | p_i \rangle^c \quad (5\text{-}103)$$

The time-ordered operator is expressed by Wick's theorem in terms of normal products involving as coefficients the contractions

$$\overline{\psi_\zeta(x) \bar{\psi}_{\zeta'}(y)} = -\overline{\bar{\psi}_{\zeta'}(y) \psi_\zeta(x)} = i S^F_{\zeta\zeta'}(x-y) = i \int \frac{d^4q}{(2\pi)^4} \, e^{-iq \cdot (x-y)} \left(\frac{1}{\not{q} - m + i\varepsilon} \right)_{\zeta\zeta'}$$

$$\overline{\psi_\zeta(x) \psi_{\zeta'}(y)} = \overline{\bar{\psi}_\zeta(x) \bar{\psi}_{\zeta'}(y)} = 0$$

Let us apply Eq. (4-75); we omit contractions between operators referring to the same point because of the normal ordering of the current. Thus

$$T : \bar{\psi}(x) \gamma_\rho \psi(x) : \; : \bar{\psi}(y) \gamma_\nu \psi(y) :$$
$$= i : \bar{\psi}(x) \gamma_\rho S^F(x-y) \gamma_\nu \psi(y) : + i : \bar{\psi}(y) \gamma_\nu S^F(y-x) \gamma_\rho \psi(x) : + \cdots$$

The remaining terms do not contribute to the connected matrix element. We expand the free fields ψ and $\bar{\psi}$ in terms of creation and annihilation operators [Eq. (3-157)]. If α_i and α_f denote the initial and final electron polarizations

$$|p_i\rangle = b^\dagger(p_i, \alpha_i) |0\rangle \qquad |p_f\rangle = b^\dagger(p_f, \alpha_f) |0\rangle$$

we have typically to evaluate

$$\langle 0 | \, b(p_f, \alpha_f) : \bar{\psi}_\zeta(x) \psi_\eta(y) : b^\dagger(p_i, \alpha_i) | 0 \rangle = \int \int \frac{d^3q_1}{(2\pi)^3} \frac{d^3q_2}{(2\pi)^3} \frac{m^2}{q_1^0 q_2^0} e^{i(q_1 \cdot x - q_2 \cdot y)}$$

$$\times \sum_{\alpha_1, \alpha_2} \bar{u}_\zeta(q_1, \alpha_1) u_\eta(q_2, \alpha_2) \langle 0 | b(p_f, \alpha_f) b^\dagger(q_1, \alpha_1) b(q_2, \alpha_2) b^\dagger(q_i, \alpha_i) | 0 \rangle$$

$$= e^{i(p_f \cdot x - p_i \cdot y)} \bar{u}_\zeta(p_f, \alpha_f) u_\eta(p_i, \alpha_i)$$

This allows us to compute $S^c_{f \leftarrow i}$ as

$$S^c_{f \leftarrow i} = -ie^2 \int \int d^4x \, d^4y \, e^{i(k_f \cdot x - k_i \cdot y)} \int \frac{d^4q}{(2\pi)^4}$$

$$\times \left[e^{i(p_f - q) \cdot x - i(p_i - q) \cdot y} \, \bar{u}(p_f, \alpha_f) \not{\epsilon}_f \, \frac{1}{\not{q} - m + i\varepsilon} \, \not{\epsilon}_i u(p_i, \alpha_i) \right.$$

$$\left. + e^{i(p_f - q) \cdot y - i(p_i - q) \cdot x} \, \bar{u}(p_f, \alpha_f) \not{\epsilon}_i \, \frac{1}{\not{q} - m + i\varepsilon} \, \not{\epsilon}_f u(p_i, \alpha_i) \right]$$

Figure 1. A page of physics.
A page of calculation in the quantum field theory, taken from a chapter entitled 'Elementary process'. One wonders what would become of this riot of abstruse symbols if the processes studied were *not* elementary (Claude Itzykson and Jean Bernard Zuber, *Quantum Field Theory*. McGraw Hill, 1980).

the sensory world and the intelligible world, which Aristotelian physics had hitherto confused. But whereas the maths of classical physics remained relatively elementary, that of quantum physics is no longer. It juggles in *Hilbert spaces* with *symmetric groups, wave functions, state vectors, proper values, density matrices* and *Hermite polynomials*. How many of us can claim to be familiar with all these things?

In this little formal game, the abstract has become a component of reality. Modern theories of physics have an undeniable propensity (some might say a vexing tendency) to distance themselves from the concrete. They contain more and more equations and teem with abstruse signs. Mathematics dictates the argument.

But this mathematisation is not just an aesthetic game. At the same time as invading physical theories, abstraction has made them more fertile. In doing so it pays implicit homage to the thought of Pythagoras (*'Everything is number'*). Is this power of abstraction evidence of a sort of kinship between nature and mind? Or has maths appeared in the field of physics only in order to endow it with a precise and unambiguous language? As these questions are obviously undecidable (by us), we shall leave them wide open.

Meanwhile, the sheer profusion of mathematics is a massive obstacle to all attempts at popularisation. Because of this, the bridges between scientist and layman are down, or rather they only take one-way traffic. It all works out as if science were building very fine, vast and elegant cathedrals, but on sites that very few persons can visit. A kind of linguistic barrier has gone up between those with and those without sufficient mathematical knowledge to understand the subtleties of science. Of course, one may deplore the fact that such a partition exists, and regret that mathematics should be so difficult, but this would not be very original, as it has never had a reputation for easy access. Legend relates that a king who had asked Euclid to teach him geometry complained of its difficulty. Euclid simply replied: 'There is no royal road.'

Galileo said that mathematics is the language of nature, and this view was no doubt more metaphysical than physical, but what is certain is that mathematics has become the grammar of physics, and that the rules of this grammar must be learned by those who wish to describe nature. There is no alternative.

One may deplore and object to this hegemonist turn of mathematics. It is true that our minds are not spontaneously drawn to unitary matrices, tensors of rank two, four-dimensional geometry or the square root of -1. The fact remains that we are no longer entitled to project our human, all too human, concepts onto the microcosm. They are unfit to be exported into the little world of particles, because the universe does not reproduce a miniature image of the infinitely great in the infinitely small. The two scales are at odds. A chasm separates the atom from the model of a planetary system that was initially proposed to represent it. The seemingly most obvious notions, those that no one ever thought of contesting a priori, begin to lose their meaning at the bounds of matter. Concepts as simple as those of position and velocity have had to be radically transformed: Heisenberg's indeterminacy principle shows that, in a quantum context, one can no longer speak of the position and the velocity of a particle, if by this one understands arbitrarily precise values possessed by both these entities at once. Thus physicists have learned to distrust and even despise elementary ideas, because they no longer express what is real.

This renunciation of familiar ideas can be a painful and sometimes touching sacrifice for the intellect, but fortunately it does not leave us entirely bereft, because reason has shown that it has something to fall back on. While pooh-poohing images, descriptions and models, it is capable of transcending the usual language. Right at the point where all these procedures break down, it has managed to base a representation of phenomena – one that allows them to be predicted – on pure mathematical symbolism. No one is capable of *seeing* in the four-dimensional space-time of special relativity. But there does exist a rigorous mathematical formalism that allows the performance of all the operations and calculations required for its description. Thus mathematics partially makes up for the defects of our senses. Thanks to it, physics has been able to transcend the contingency that presided at its birth. It has expanded. Only mathematics can transport us beyond the specific conditions of our immediate environment.

Because of the atom and the photon, simplicity ran aground, and its disappearance did not make everyone happy. Max Planck, who discovered the quantum in 1900, was very distressed by it. He did not look kindly on the changes his discoveries had provoked. It was he who reluctantly introduced, to resolve the black-body radiation problem, the famous constant h, which contains in embryo the entire quantum revolution.

The black-body problem was as follows. Every piece of matter, when heated, begins to radiate, passing from red to white heat as its temperature rises. This change of colour depends hardly at all on the nature of the surface, and for a black body the colour depends only on the temperature. Therefore the radiation emitted by this black body at high temperatures is a simple phenomenon which should admit of a simple explanation by the known laws of radiation and heat. However, the application of these laws led to a result that did not tally with the experimental findings.

Planck attacked the problem in 1895. He tried to represent the measurements of the spectrum of thermal radiation by simple mathematical equations, and found one that gave a perfect account of the most precise experimental results. Now he embarked on an intense theoretical study. What could explain the new equation? He discovered that the whole process worked as if the atoms radiating from the surface of the black body could only exchange energy in the form of little packets of energy that he called *quanta*. (This word is the Latin plural of *quantum*, which means 'how much', and which also appears in 'quantity'.) His finding differed so much from what was then known in classical physics that Planck had to reject it at the outset, but eventually he convinced himself, during the summer of 1900, that the conclusion was inescapable. He published his hypothesis of the quantum in December 1900, and introduced the tiny constant h (equal to 6.63×10^{-34} joule second) that bears his name.

The existence of this constant embarrassed Planck because he felt that it threw doubt on the very foundations of classical physics, since it injected discontinuity into processes which had always been considered as fundamentally continuous. In the years following the publication of his article, he did his utmost to reinterpret his hypothesis to accord with the old systems, but failed on the essential points. A good loser, he recognised that one could no longer hope to describe the world of microphysics with the tools of classical physics. New paths were opening for physics, which would lead it permanently away from its former, too ordinary, perspectives. In his *Introduction to Physics* Max Planck wrote: 'This movement by virtue of which the physical understanding of the universe departs further and further from the sensible world the more it is perfected, is also a movement of closer and closer approach towards the real world.'

None of the recent developments in physics requires any change in this assertion; far from it. Let us take the example of the *Pauli exclusion principle*, which asserts that no two electrons in an atom can exist in the same quantum state. This principle flows from a link between the symmetry of the state vector of a system of several identical particles and the spin of these particles. This link was originally an experimental fact. Recently it has become a theorem, because it can be demonstrated. But to do this, one must resort to a kind of theoretical juggernaut, the relativistic quantum field theory. Its use calls for courage, perseverance and nerve, but there is no other tool with which to prove the theorem in question. So there exist in physics simple laws that are simple to state, and demonstrable within the frame of current theories, but which cannot be simply explained. Even Richard Feynman, Nobel prizewinner for physics in 1965 and a great teacher if there ever was one, admitted that he did not feel up to it.

To fit them to describe reality, we have to make the effort to expand our conceptions until they are unrecognisable. This is exactly the price that the quantum pioneers had to pay. The successes of modern mathematical physics prove that this investment was a profitable one, especially because maths is no longer a purely servile language. It has turned out to have its own autonomy which can make it fertile in other disciplines. IN 1623, Francis Bacon had noted the first signs of this evolution, but that was in order to deplore it. He saw it mainly as a form of culpable arrogance that had to be opposed. In *The Advancement of Learning* he admitted that he did not understand 'how it happens that logic and mathematics, which should be only the servants of physics, sometimes priding themselves on their certitudes, wish absolutely to lay down its laws.'

In reality, the major mathematical discoveries have done much more than suggest ideas useful to the physicist. The spectacular generative power of mathematics has come into play in a forcible and probably irreversible fashion. This really began with the great Scottish physicist, James Clerk Maxwell, famous for his equations. When he started working, the laws of electromagnetism then accepted took account of all the experimental facts, but by viewing them from a new perspective, Maxwell realised that the equations became more symmetrical, and therefore more beautiful, when one term was added. As this term proved to be too small to produce appreciable differences from the old methods, its omission had hitherto had no serious implications. These intuitions of Maxwell's, which anticipated among other things the existence of electromagnetic waves, had to wait twenty years for experimental confirmation, thanks to the work of Hertz. In other words, because he was permeated by a feeling for symmetry in mathematics, Maxwell anticipated experiment by twenty years. Still, this was no less a gain.

Mathematics is implicit in the field of physics. It is here that the 'true creative principle' resides, to use Einstein's words. It has happened several times that speculations of a purely mathematical nature have given rise to objects entirely new in physics. This was the case with the antiparticle of the electron, called the *positron*, which was 'calculated' by Dirac in 1927 before being discovered in 1932 in cosmic radiation. The positron had been postulated by Dirac as one of the solutions of the new equation he had constructed on the basis of the idea that the formalisms arising from quantum mechanics had to allow for the postulates of special relativity.

The contribution of mathematics culminates in physics in the context of the so-called 'gauge theories', which are an elaborated form of the quantum field theory. The oldest gauge theory is that of electromagnetism. Anyone who has learned a little physics will know about the 'potentials' of electromagnetism, on the basis of which electrical and magnetic fields may be calculated. They will also know that these potentials are not directly observable as such, since Maxwell's equations do not determine the potentials themselves in absolute fashion, but only consider potential differences. In particular, the reference framework of these potentials may be modified arbitrarily without overturning the laws of physics. These modifications of potential are called 'gauge transformations', and Maxwell's equations are called 'gauge invariants'. Later, in the 1950s, the method of utilising the gauge theories was formulated and extended to other types of interactions by two theorists, C. N. Yang and R. Mills. These theories employ the language of mathematics, since the basic or Lagrangian notion they employ makes sense only within the framework of the integral calculus. A feature of these theories is that, starting with a fundamental principle of invariance and symmetry, they lead to the necessity to introduce new entities into physics, for example new particles. They are therefore capable of enriching the ontology of physics. In particular, it is with their help that physicists hope to unify all the fundamental interactions of nature.

The gauge theories already have great successes to their credit, such as that of the unification of the electromagnetic and weak interactions, which has given substance to the concept of *electroweak interaction* and earned the Nobel Prize in 1979 for three theoreticians: Sheldon Glashow, Abdus Salam and Steven Weinberg. The development of these theories, complemented by the introduction of the concept of 'spontaneous symmetry breaking', led to the prediction of the existence of three new massive particles, one neutral and two charged, called the intermediate vector bosons (Z^0, W^+ and W^-), which constitute a sort of heavy light. These new particles were in fact demonstrated in 1983 at CERN, during high-energy proton-antiproton collisions.

Long gone are the days when mathematics could be reduced to the level of a mere tool in the service of physics. It now has the power to extend physics beyond its field of experiment, enrich its range of concepts and increase the number of its objects. In structuring the discourse that physicists conduct about reality, mathematics provides physics with the means to grow sharper, and cut deeper.

8

Genius, or the Art of Being Naïve

*You see things; and you say 'Why?' But I dream
things that never were; and I say 'Why not?'*
Bernard Shaw

Someone once said that no scientific mind can call itself really young, since it is at least as old as its prejudices. There is no absolute freshness, no neutral knowledge, no achromatic eye. Each fact acquired infuses its possessor. But, faced with a surprise or a novelty he does not understand, the physicist must revert to a kind of naïvity that frees him from the automatisms of everyday experience. In order to be ready to envisage, to doubt, and to invent, he must first try to regain an heretical freshness of approach, an argued ingenuousness, in other words a beginner's state of mind.

Because confidence in a discovery is not in itself a proof, physics has constructed its own criteria of validation, which accept only what is confirmed by experiment. Neither logical consistency nor mathematical elegance suffices in itself to back a physical theory, even if they are often valuable indicators. Conversely, a thesis founded on experiment may not be rejected simply because of its alleged unsoundness. In 1905, it was established that light was made up of electromagnetic waves. Young, Fresnel, Maxwell and some others had demonstrated it officially, and brilliantly.

The wave theory, confirmed by experiment, had become canonical. Despite this, Einstein had the nerve to wake the Sleeping Beauty of Newton's old corpuscular theory so as to account for the photoelectric effect – the emission of electrons by metals under the influence of light. Experiments had shown that the energy of the electrons emitted did not depend on the intensity of the light employed, but solely on its colour or, more precisely, its frequency. This could not be understood in terms of the classical theory of radiation.

Einstein claimed to explain these observations by applying Planck's quantum theory. Instead of representing light energy as a continuous flow, he supposed that this energy is emitted in discontinuous fashion, in *quanta*. In so far as we may use metaphors here, the image he gave of light was not that of an evenly flowing river but of a sporadic flow of water in which a dam spills over all at once whenever its content reaches a certain level. Thus Einstein transmitted the quantum disease to light itself, and because of that was challenged by Planck himself, the father of the quantum, who felt that classical physics had suffered damage enough already! Einstein was unruffled. He showed that, in order to accord with Planck's hypotheses, the energy of a light quantum must be equal to the frequency of the light multiplied by Planck's constant h. In writing this, he attributed a corpuscular structure to radiation itself, and no longer simply to energy exchanges, each light corpuscle transporting a quantum of energy. He was of course aware of the phenomena of the diffraction and interference of light, and knew perfectly well that they could only be explained by the wave model, i.e. that of an even wave flowing without a break. Hence he could not deny that there was a contradiction between the wave model and the idea of the quantum of light that he was using. But since certain experiments could not be explained except by the corpuscular interpretation, he did not try to reject it. He accepted the contradiction as something that would probably be understood later on, but which allowed progress for now. This is exactly what happened. Einstein's theses led Louis de Broglie to the concept of wave-particle duality, which is at the base of all quantum physics.

In introducing this concept of wave-particle duality, de Broglie also gave proof of freshness of approach. He showed mathematically that every wave is equivalent to a particle, and that, reciprocally, every particle can be interpreted in terms of waves. Before him, physicists considered that it was impossible for a physical object to be at once a wave and a particle. How could science conceptually combine the infinite and the point? The two definitions – *the electron is a wave* and *the electron is a particle* – seemed mutually exclusive. They had the same subject, but predicates that contradicted each other as sharply as fire and water.

To avoid this contradiction de Broglie resolved to consider things from a new angle. He devised the extraordinary equation $\lambda = h/p$, which links a pure wave magnitude (the wavelength λ) to a pure mechanical magnitude (the momentum p). This equation allowed the switch from an exclusive to a relational opposition. The starkness of the wave-particle confrontation faded into the light and shade of a dream of peace, now called the *wave function*.

The true scientist is the one whose mind stops at things that do not strike other people, and by dwelling there manages to look at what others do not see. The history of the theory of relativity is very eloquent from this point of view. For several centuries, scientific thought lived with the idea that time was absolute, and independent of the observer. This (Newtonian) conception of time made it possible to postulate the existence of a class of special reference frames, called Galilean, in which all the laws of physics were identical when one moved from one frame to another. After Maxwell had put the electromagnetic waves into equations, the physicists thought it necessary to imagine the existence of a substance within which these waves were propagated. Why was it not possible to believe, at that time, that waves might be propagated in a vacuum? It was known that when a bell was rung in a box containing a vacuum, no sound reached the exterior because acoustic waves need air to propagate. Because it was known, on the other hand, that light will cross a vacuum without difficulty, it was supposed that light waves might be considered as elastic waves in a very light substance that was given the name of *ether*. This invisible, impalpable substance was assumed to fill both the vacuum container and the space in which there were other substances like air or glass. The idea that electromagnetic waves might have a reality in themselves, independent of any substance, was unthinkable for the physicists of that epoch. But they asked a host of questions on the subject of this hypothetical ether. Since it seemed to penetrate everywhere, including the interior of matter, they wondered what happened when this matter was set in motion. Did the ether too participate in the movement? And, if so, how could a light wave be propagated in moving ether? Many physicists thought that the ether could only be the natural anchorage point of a privileged reference frame, representing absolute immobility, and by reference to which it would be possible to define absolute movement.

All classical mechanics, as well as the mechanistic interpretation of electromagnetism, was based on these principles. Yet no one had ever succeeded in finding a strictly Galilean reference framework, and the experiment by Michelson and Morley, which was designed to measure the speed of the earth in relation to the ether, had ended in failure (see the appendix at the end of this chapter p. 70). Against all expectations, it indicated that the speed of light in relation to the earth was the same in all directions, which signified that the earth was motionless in relation to the ether. But it was known that this was impossible, since the reference frames linked to the earth are not Galilean (because of the presence of forces of inertia). No satisfactory interpretation was advanced to explain these paradoxical results before 1905, the date at which Einstein published his special theory of relativity. Michelson's experiment indicated that the speed of light in a vacuum is constant in relation to any reference frame. As a good creative rebel, Einstein raised this paradox into a postulate, and as this postulate was contradictory to Galileo's law of composition of speeds, he adopted new rules of transformation, borrowed from his former mathematics teacher, Henrik Lorentz. He added that, since no one had ever managed to lay a finger on the ether, then it simply must not exist.

Therefore electromagnetic waves were propagated in a vacuum! He also concluded that if you found that you were measuring the same speed of light propagation whether you were in motion or at rest, it was because the ruler with which you made the measurement was shorter when it was in motion and the clock of measurement was slower. There was a *contraction of lengths* and a *dilation of durations*.

Thus, instead of feeling passive surprise that experiment contradicted theory, Einstein's bold reflex was to modify the very basis of mechanics, something that physicists had been refusing to do for eighteen years. (It is true that the modification was quite radical.) He was the first to dare to pronounce the death sentence on the theory of a motionless ocean of ether. It was by discarding this unwanted quest of classical physics that he was able to construct his theory of relativity, which is now a mainstay of the whole of physics.

In physics, as elsewhere, daring may have its triumphs.

The Michelson-Morley experiment, or the ether at bay

In 1881 the American physicist Albert Michelson wanted to define and measure the 'ethereal wind' that was bound to be produced by the movement of the earth in space. To do this, he had the idea of using the phenomenon of interference of light waves: two waves that interfere produce a pattern consisting of a luminous point surrounded by a series of alternately dark and bright rings, depending on whether their phase differences equal a half-wave length or an entire multiple of the wavelength. If one of the two paths is changed, even very slightly, the interference figure is shifted sideways. Thus, by means of interference, the distance travelled by the light during a certain time can be measured. Michelson took advantage of this brilliant idea to develop an ultrasensitive instrument, now known as the Michelson interferometer, which would make it possible to demonstrate the way the movement of the earth modified the speed of light as measured in the laboratory.

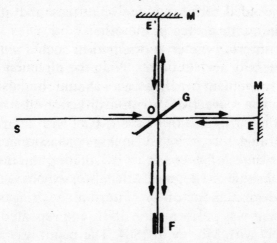

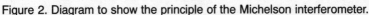

Figure 2. Diagram to show the principle of the Michelson interferometer.

This interferometer is outlined in Figure 2. A light source S delivers a beam SO which arrives at a semi-transparent mirror placed at G, where it is split in two. One part of the incident beam continues on its way (OE); the rest is reflected (OE'). The two rays propagated along OE and OE' are reflected by the mirrors M and M' and return along their direction of incidence. Once more, the semitransparent mirror placed at G reflects them and partly transmits them. E'O is transmitted as OF while EO is reflected as OF. Along OF there are therefore propagated two rays issuing from the same source but which have followed different optical paths; they interfere. The least variation in one of the optical pathways produces a locatable shift in the interference rings. In particular, if, as was thought, the earth's movement modified the lengths of the optic pathways, a simple pivoting of the apparatus should produce a shift in the interference patterns. The orientations of the apparatus that led to the greatest shifts should determine along what axis and at what speed the earth moves in the ether.

If the interferometer is oriented in such a way that the incident beam SO points along the direction of the earth's movement, interference should be observed between the wave (MO) with a direction identical to that of the earth and the wave (MO') with a direction at right angles to this. By making measurements at different times of the year, or by rotating the interferometer through 90 degrees, Michelson expected to witness a displacement of the interference fringes. The experiment was performed in 1881, and repeated more accurately with Morley in 1884. The result was always negative. Michelson and Morley observed no apparent variation of the speed of light when the movement of the interferometer in relation to the ether changed direction. This finding utterly baffled all physicists. If the earth travelled in the ether, how was it that it was impossible to demonstrate the wind it created? Michelson published his conclusions, but he was very disappointed. He had pinned his hopes on becoming the first to detect the movement of the earth through the ether. This negative result was so embarrassing that the experiment was repeated again and again, by Michelson and Morley themselves, as well as by others, but the findings were always the same. This meant that physicists were now obliged to look hard at the problem of the ether, and to confront the two following paradoxes:

1 The ether exists, but the movement of the earth does not produce an ethereal wind.
2 The velocity of light measured from a moving earth is always constant!

As regards the ether, the physicists were impaled on the horns of a dilemma. Either they abandoned the concept of this altogether too imaginary environment – but then they could no longer explain the propagation of electromagnetic waves; or else they continued to assume the existence of the ether – but then they would have to reject the theses of Copernicus and admit the immobility of the earth! Michelson, for his part, was prepared to circumvent the problem by saying that all the ether in the universe was dragged along by the earth, which amounted to restoring the earth to a privileged position and role in the universe. This dilemma embarrassed everyone. The Irish physicist George Fitzgerald proposed the following explanation: the failure of Michelson's experiment could be understood if it was assumed that any object moving through the ether contracted in the direction of its displacement, and increasingly so as its speed approached that of light. The shrinkage of objects undergoing displacement around us would obviously be negligible and undetectable. Most scientists greeted his idea with frank hilarity. But in 1892 the Dutch physicist Hendrik Anton Lorentz arrived independently at the same conclusion and published it. Since he was considered at the time as the great expert on electromagnetic theory, he was taken more seriously than Fitzgerald. Lorentz was able to calculate the coefficient of contraction that needed to be applied to allow for the invariance of the speed of light. The twofold problem posed by Michelson's results now seemed settled, but other difficulties emerged, for at the end of Lorentz's calculation the ether found itself divested of all its mechanical properties. It became more and more 'ethereal'.

But if the ether no longer had any mechanical properties, the question of whether it existed at all must eventually be posed. It was posed in 1905 by Einstein, who replied in the negative. So ended the career of the ether, which had lasted for more than two centuries.

9

Paradox as Demarcation Line

> It is certainly not the slightest charm of a
> theory to be refutable.
> Nietzsche, *Beyond Good and Evil*

All science starts from the principle that nature does not do just anything at all, or work just anyhow. It stands on its belief in at least some degree of intelligibility in the universe, which it seeks to express as well as it can. Since, as Descartes remarked, 'our thought imposes no necessity on things', we have to abandon the idea of alighting on the real only by exploring the mind. A theory that has claims to scientific status must be put to the test of experience. To make this feasible, there has to be a power to make predictions, which sets science apart from writing essays or collecting stamps.

Thanks to experiment, scientists have the means to give a voice to the real, or to what, for them, takes its place. In seeking to know what answers it gives to their questions, they may happen on a result that tells them: Non, niet, no, nein! No, Newtonian attraction does not explain the advance of the perihelion of Mercury; no, the universe is not static; no, the atom is not indivisible; no, the particle is not a small body; no, quarks are not observable in isolation. Paradoxically, these negative answers plays a positive role, because by proving established science wrong they move trail-blazing science forward. The French epistemologist Gaston Bachelard devoted a fine book, *The Philosophy of the No*, to this incessant polemic activity that pervades the very foundations of science. If it takes the shape of a paradox or an anomaly, each negative answer from the real leads scientists to dream up new hypotheses and embark on new paths.

Thus the physicist is constantly required to draw the Leibnizian distinction between possible worlds and the real world. In a zigzag sequence it sets new questions for the physical world, puts them into shape, goes looking for the answer, devises explanations, formulates other questions, tracks down new answers, and so on and on. Very schematically, it is just like this that science evolves. Galileo's '*Eppur si muove*' therefore applies to science itself. It too keeps moving.

Reality is interpreted in the rhythmic swing of the oscillations between idea and thing, shuttling between the concrete and the abstract. To do science means thinking that description and explanation are two movements that converge upon each other, even at an infinite point; it means imagining that the meeting of the two, even though perpetually receding, lies somewhere at the end of the asymptote, and each advance of knowledge brings it closer. Science is a straining towards precision, a tension that never relaxes. Hence the question of the distance to the asymptote remains open. Erwin Schrödinger considered, in *Mind and Matter*, that it could well be that in every new advance of science, 'the absentee glimpsed out of the corner of the eye remains forever nameless'.

One may wonder too whether the order revealed by physics shows the face of a geometrician god, a lover of mathematics, or whether it is the one comprehensible island in an ocean of disorder. This question brings many others in its wake, all of them metaphysical. What is there missing from reality, that stops it being instantly recognised by ourselves? Does it lack transparency, or is it our own intelligence that wears permanent shackles, and is condemned never to know everything? May we hope some day to gain a firm grip on the world? But then again, will we be able to tell the difference between the provinces of human constructs and of observable reality? These questions, which riddle the entire history of philosophy, are colossal issues, and possibly even undecidable. Since our own lucid lack of expertise deters us from replying, we shall not mention them again, except to stress that paradoxes function at the heart of these strenuous vacillations between reality and its portrayal. They intervene to exhibit a braying dysharmony, a perceptible friction between object and discourse. But in order to emerge, they need a little space next door (*para*) to knowledge (*doxa*), as their etymology proclaims. So long as physics makes no precise and faithful match with the contours of reality, and so long as its laws are not the laws of nature, its predictions will never be immune from being refuted. Somewhere between reality and arithmetic, paradoxes might eventually find a little place to settle, and from there to open fire.

Scientific knowledge is therefore incapable in principle of distilling into a worthy, definitive, ready-to-think concoction. All the propositions of science, even the most famous, are only moments and agencies in an infinite task. Infinite, because even if science should one day reach completion, even if everything seemed to indicate that it had run its course, it would lack the wherewithal to say so of itself. Stopping does not mean arriving, as mountaineers are well aware. So it would not be enough for all the problems posed to have been solved, before it could be said that the task of science was over and done. It would also require all possible problems to have actually been posed – a verdict of which it is impossible to be certain. An unexpected fact, a new experiment, or an improvement in the accuracy of measurements, might detect gaps or flaws in an area that the physicists of the time had considered totally safe and sound, and so revive the march of science. None of our descriptions is imperishable, none of our models unchallengeable, no theory can be guaranteed for ever.

Think for example of the various atomic schemas contrived by physicists over the past hundred years. It is common knowledge that the atom consists of a nucleus, positively charged, surrounded by negatively charged electrons. Among the first descriptions made of it was the one put forward by Niels Bohr in 1913 (see Figure 3). It likened the atom to a miniature planetary system (the electrons orbiting the nucleus) of a very special kind, since the orbits were not free to alter. It was very soon realised that this description had better not be taken too literally. The 'orbits' of the electrons proved hard to represent by a real motion in space, as if the ordinary notion of trajectory ceased to apply inside the atom. It also proved necessary to refrain from exploring the characters of suddenness implied by the *quantum jump* idea (when the electron moves from one orbit to another), because the elec-

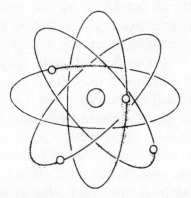

Figure 3. Bohr's model of the atom.
The Rutherford-Bohr atomic model, which has become the symbol of
the American Atomic Energy Commission. The electrons orbit the central nucleus rather as the planets of the solar system orbit the sun.

tron seems not to be localised in the atom in the
manner suggested by this image. A brief summary of
the story might go as follows. At first the physicists
drew up a detailed plan of the atom, and then their
critical spirit stepped in to compel them to omit certain
details, and sometimes a lot more than details. What
survived this bout of iconoclasm was very exactly the
atom of modern physics: a diagram of probabilities
(known as a *matrix*) which has very little to do with
the initial planetary system. As today's atom is precisely the sum of the criticisms levelled at its prototype,
it cannot be understood without reference to the history of its portrayal, which shows that even if the
schema devised by Bohr worked well as an image,
hardly anything remains of his original version. Not
that it was labour in vain, for this is just the way that
great ideas are born, and it is because it has obstacles
to surmount that science can build a considered train
of thought. If we knew all the ideas that science has

thus discarded, we would be all the more impressed by those it has preserved. The selection system that it operates is beautifully, inflexibly, severe.

As science is (in principle) methodical in its progress, it can be assigned a purposeful plan, which is to convert the ground of the obstacles it tackles into the site of a clean break between rigorous thinking and the random travels of pseudo-science, dogma and ideology. Are there other ways to distinguish true science from the pseudo-sciences? When does science begin, where does it end, and what is its signature? What does the scientificness of science depend upon? What entitles a given discourse to be called *scientific*? These are fearsome questions, to which – paradoxically – there are no scientific answers. However, in his book *The Logic of Scientific Discovery* (1935), Karl Popper attempted a reply.

He began by showing that it was no use clinging to the classic (Baconian) thesis which held that the sciences were characterised by their inductive method, i.e. that they were based on reasoning that depended on moving from particular statements to general statements, laws or theories ('I have never seen anything but white swans, therefore all swans are white'). In fact, Popper points out, they are not at all the outcome of a process of generalising from individual cases. Furthermore, induction cannot be considered as a rigorous reasoning process, because what is true of the individual or the part is not necessarily true of the whole: the whiteness of the countless swans I have been able to observe does not prove that *all* swans are white, and the next one I see may just as well be black. David Hume, in the eighteenth century, had already provided a vehement critique of induction, applying it more particularly to the notion of causality. 'We have no arguments,' he wrote in his *Enquiry Concerning Human Understanding*, 'to convince us that the objects which, in our experience, have been frequently conjoined, will be conjoined in the same manner in other circumstances.' This means that in respect to the empirical domain to be explored, every theory represents a reasoning a priori. Far from being the passive recording of given regularities, the formation of a hypothesis is an active, creative exercise. We do not argue from facts to theories, except by way of refutation. Karl Popper goes so far as to write that induction – and hence the verifiability of a scientific thesis – is a myth. It cannot be set up either as logic or as method, because it reaches only hypothetical conclusions, which amount to no conclusions at all.

Popper goes on to make a cardinal assertion, based on considerations of elementary logic. Whereas it is false that a general, universal proposition can be decisively proved on the basis of singular or particular statements (because what goes for the individual or the part does not necessarily go for the whole), on the other hand it is true that a universal statement can be falsified – correctly classed as false: all that takes is for at least one case to contradict the universal statement. In axiomatic language, this comes down to saying that what is denied in the case of an individual or group of individuals cannot or can no longer be affirmed about the whole. Let us take a very simple example, far removed from physics (an occasional change of scenery is healthy). When thinking about mammals, there is a temptation (before the discovery of Australia an effective temptation) to decree that all the animals in this class are viviparous. This universal statement was categorically falsified, i.e. contradicted, by the existence and discovery of the platypus (in French *ornithorynque paradoxal!*), which is not viviparous but none the less a mammal. Another example, this time from mycology: around the middle of this century it was generally agreed that all *Cortinarius* species could be eaten with perfect safety, because no poisonous kind had ever been found. Until the day came when this prejudice was exploded by the alarming discovery of the species *Cortinarius orellanus*, which caused a serious outbreak of poisoning in Poland. So dogmatism can have fatal consequences (not only for mushroom lovers) that are well worth dwelling upon.

Popper therefore proposed to replace induction by what he called the *testability* or *falsifiability* of hypotheses, which he made the true touchstone of the scientificness of statements, in accordance with a criterion which is dually negative: 'A theory that is not refutable is not a scientific theory,' Popper explains. Consequently, the scientific attitude becomes a critical attitude that looks for crucial tests, i.e. tests that may refute the theory, but can never establish it for sure. This assumes on the one hand that it is predictive, on the other that it has points of contact with reality, where it can be tested. It is at the time of these contacts with the real that the limits of the hypothesis appear and that possible paradoxes arise. In other words, if Popper is to be believed, a theory is scientific not because it is true, but because it enables its own faults and lacks to be demonstrated ('The only thinking that lives,' said Dostoevsky, 'is the kind that maintains itself at the temperature of its own destruction'). It has to be liable to be put at risk with every new and decisive experiment. A scientific statement is therefore essentially vulnerable.

At first sight, this new epistemological conception according to which a theory can only be refuted, and never empirically established, may seem very unwelcome, because it prohibits talking about truth. Must the scholar be confined to negative certitudes such as 'It isn't true that ... it is false to say that ...'? Popper is well aware of this retreat, but he makes a sound case for judging his criterion preferable to the rest: he considers it more in keeping with the way the story of scientific discovery really unfolds.

So what now emerges is a marked difference between closed doctrines that reject all commerce with the outside world in case it should damage or diminish them, and open theories that run the intellectual risk of their own transformation. Closed theories being irrefutable dogmas, only open theories can be considered scientific. That is the Popperian criterion of the scientificness of a discourse. In this schema, the scientist is no longer quite the scholar. He is not the one who knows, but the one who can ask critical and provocative questions.

This criterion, like all criteria, is certainly not perfect. Its importance derives, in our view, from its implicit recognition of the fundamental role of paradoxes. In fact to consider that theories are scientific only if they are capable of admitting their own negation amounts to measuring their scientificness against the standard of their capacity to become paradoxical. In other words, this is grist to our thematic mill. After all, Hegel must have had much the same thing in mind when he wrote that 'knowledge does not know itself only as itself, but also as its opposite or limit'. But that is enough interpretation.

Curiously, paradoxes reflect both the unfinished state of science and also its degree of maturity, which is what one might call the paradox of paradox! A finished theory, one that was by definition always correct, could not be paradoxical. Paradoxes are therefore the sign of incompletion, imperfection. But conversely, only a mature and major theory, rich in content, can say enough to be possibly refuted by the facts. In fact paradoxes can only appear if the theories that contain them are pointed, precise and sharp enough to present a cutting edge when they come to confront reality.

Some theories are toffees, their shape so malleable that they mould themselves obediently to every form of reality. What type of facts would have the power to refute, or even to object to them? They are structurally unassailable by experiment. In particular, if a theory is formulated in terms so vague that it is impossible to tell quite what it says, then no matter what tests it undergoes it can always be interpreted so as to comply with their findings. In his *Treatise on Colours*, Goethe wrote about electricity in the early nineteenth century that: 'It is a nothingness, a zero, a point zero, a neutral point, but one that is found in all beings and is at the same time the source that produces, at the slightest opportunity, a dual phenomenon that appears only to disappear again. The conditions that determine this appearance are infinitely different according to the constitution of the particular bodies.' If we took this quotation literally, it would be very hard to find circumstances in which it might be falsified, since it blandly asserts both everything and its opposite. Not being predictive, it runs no risk of being refuted.

This comment applies to more than just the sciences. In his book *The Language of Cotton*, François-Bernard Huyghe pointed out that diplomats and other politicians are increasingly using a watered-down language whose few and hence inflated words no longer have any true meaning; a consummate consensual language that panders to the taste for tautology and disables contradiction; a discourse which has an answer to everything because it says practically nothing; a language unanswerable because it churns out propositions that leave so much room for interpretation that listeners are free to hear what they hope for. In other words, a language so all-inclusive that it gives no chance to paradox – and here there are grounds for unease, we must confess. Who could argue with a minister who learnedly proclaims (as one has done): 'We must reaffirm permanent values and invent a new way of thinking'? Who could contest the ecumenical wisdom of such a remark? No one, of course. The trouble is that this message, which sits so neatly on the fence, is not altogether a message, since it cannot be contradicted. Hence one may wonder whether it still has a meaning at all.

Coming back to physics, the point to be reiterated is that spongy theories, able to soak up everything, cannot be refuted. That is their scientific sin. Since nothing can trigger their development, they have no dynamic. They are capable of living, or anyway surviving, eternally without a change of composition. Alfred North Whitehead said that 'theology is less changeable than science', and it is fair to say that theology cannot claim to be scientific because it owes its great stability to the fact that it is founded on an unverifiable world. Its status is neither higher nor lower than the status of the sciences; it is simply other. Science inheres in the open character of the venture that permits the continued questioning of its own structures of thought. It is by its capacity to collide with itself that it becomes science.

A theory that does not commit itself has no cause to change. In accommodating itself to everything, it marvels at nothing. The concept of paradox is foreign to it. It's own time is locked up, like that of the photon. Does this observation entitle us to conclude that paradoxes, because they release us from the temptation of dogma, are one of the signatures of scientificness? The urge to do so is very strong, since paradoxes are implicitly involved, as we have seen, in the criterion that demarcates science from the pseudo-sciences. From there to suggesting that a theory is scientific only on condition that it is paradox-productive is only a single step that, were it not for our timidity, we would be delighted to take.

This is the moment to recall Cioran's remark that: 'Religions die for lack of paradoxes.' The corollary of this statement is that; 'Sciences live only by virtue of paradoxes.' Well then, it seems that it really is to the heroes of this book that the sciences owe their dynamism and liveliness. Then banish timidity. Let us proclaim at long last, and unblushingly, that the vital principle of science is paradox.

10

Paradox, or the Tempo of Science

> Routine is the preface to revolution.
> *Emile de Girardin*

Science has a sense of rhythm. It is a perpetual becoming, punctuated by the shock of the antagonisms that fertilise it. The system by which it evolves is not constant, or even regular. Neither is it totally chaotic. Rather, it is governed by a combination of active conservation and upheaval. The resulting evolution is Darwinian, but of a non-gradualist kind, since it manages to alternate periods of almost flat calm with explosive episodes. The former may last long. In retrospect there is only one step from Archimedes to Galileo, or from Aristarchus of Samos, the true father of the heliocentric system, to Copernicus. But to make this step took nearly two thousand years, during which science developed rather at leisure, a little as if it had been in hibernation.

The explosive periods are easier to identify, as their name suggests. In fact, the history of the development of physics clearly shows the essential role that all kinds of breaches have played. They cause theories to succeed and supplant each other without disrupting the enterprise of science. For even when it is rocked on its foundations, the status of science is not threatened. The machine may sometimes cut out, but it does so without damaging its gears. There is no collapse. As science can contain its own divisions, its revolutions are not made against itself. A scientific revolution is *first and foremost* scientific.

The idea of scientific revolution is quite old. Lavoisier was the first to allude to it, in the eighteenth century. A little later on, the French scholar Antoine Cournot worked out its first epistemology. But it was Thomas Kuhn's book, *The Structure of Scientific Revolutions*, that threw new light on this question in 1960. In this study Kuhn introduced the concept of a *steady state* in science. According to him, science stabilises at certain periods by conforming to what he calls a *paradigm*, which defines the *normal* science within which the formulation of problems and the type of solution used are imposed in consensual fashion.

The paradigm behaves in the scientific community like a kind of shared assumption of the obvious, which very few people seek to question. You don't overturn it for fun. It is only when its anomalies or paradoxes can no longer be ignored that the investigations can begin which lead to a new basis for the practice of science. It is these extraordinary episodes, during whose course the convictions of scientists are modified and reorganised, that Kuhn calls scientific revolutions. Within these, the critical debate that in principle characterises science takes on a particularly intense and spectacular form.

Scientific revolutions are generally accompanied by terrible doubts and deep unease, because they suddenly transport scientists communally towards another conceptual planet, previously unknown. Kuhn cites, for instance, the anguished reply of the German physicist Wolfgang Pauli to what he regarded as an increasingly grave crisis in physics around 1924. The quantum theory was then a jumble of confused hypotheses, bold principles, abstruse theorems and recipes for anarchic calculations, rather than a coherent theory. Discouraged, Pauli confided to a friend: 'At the moment physics is again terribly confused. In any case, it is too difficult for me, and I wish I had been a movie comedian or something of that sort and had never heard of physics.' Happily some months later Werner Heisenberg came to the rescue of clarity in physics by proposing a mathematical structure adaptable to quantum phenomena as a whole. Pauli uttered a sigh of relief and continued his vocation as a physicist, with a Nobel Prize in 1945 into the bargain.

Scientific revolutions break the norm. They oblige the old system to give way to another, considered as better or broader. In so doing, they affect all current a priori arguments in knowledge, and compel it to reorganise itself. The most famous examples of paradigm shifts in physics include the shift from geocentrism to heliocentrism (Copernicus, 1543), the transition from the particle optics of Newton to the wave optics of Young and Fresnel, at the start of the nineteenth century, the slow birth of quantum physics (early twentieth century), special relativity (1905) and general relativity (1916). These examples are generally familiar, at least by name, and this is no doubt because major scientific revolutions are fairly few. Fortunately, scientists do not make revolutions every day. Besides, a science that was constantly busy revolutionising itself would find it hard to say what it contained, and to achieve a status that made it credible.

Certain authors have contested the idea of a normal state of science, arguing that it is in permanent flux and never calm. In every age, controversies and fluctuations disturb the scientific world, and it is rare for the hegemony of a paradigm to be so absolute that it can stifle all debate. However, it is difficult not to acknowledge the presence, in certain periods, of an overall unity of thinking about the manner of posing and resolving problems. That is the case, for example, with present-day particle physics, which leans successfully on a theory of interactions called the *standard model*. The name alone suggests the presence of a solid paradigm. So one cannot reject the idea that there are in science periods of massive consensus, not to say conformism or fashion, and that in these periods paradigms play a part analogous to that of a *doxa*. On the other hand, Kuhn may legitimately be reproached for his somewhat over-schematic view of scientific revolutions, which he describes as utter collapses, claiming that the old paradigm is entirely set aside and replaced by another. Kuhn writes as if nothing remains of the old construct, as if each revolution proclaimed: *We rub out everything and start again.* But in reality the refuted theory never dies completely. Experimental results, empirical laws, equations and concepts are reprised in the new theory in the guise of particular cases or approximations. For example, classical mechanics can be formally identified as the limit of quantum mechanics if one forgets the existence of Planck's constant. In this context it is perfectly suitable for engineers who build bridges, and it would be ridiculous to abandon it. Obviously if these same engineers continued to rely on it when it came to calculating the spectrum of the hydrogen atom, they would be gravely mistaken, because atoms cannot be described by the classical equations. Similarly, while recognising that the conceptions of Einstein and Newton differ enough to make it impossible to say that they maintain logical relations, it can be shown that Newton's theory is applicable in a great many circumstances, for instance when describing the fall of any apple or the trajectory of any satellite. But it works less well in the vicinity of a black hole, where Einstein and his general relativity reign unchallenged.

It may also happen that theories are modified without losing their validity. The theories of the 1920s, whether concerning quantum theory or relativity, have remained perfectly correct while having also been considerably modified. Actually, it is mainly our way of understanding them that has changed.

Theories change, or are changed, but facts remain. They are notoriously obstinate. From this angle (but only from this one) it may be said that Aristotelian physics remains perfectly correct, since it holds true for as long as fire rises and heavy bodies fall. Or we can take a more modern example, that of Clerk Maxwell's equations. Maxwell based his theory of electromagnetic waves on the notion of the ether, which was the medium in which these waves were supposed to propagate. Since then, the great paradigm of the ether has had to give way. We no longer believe in its existence. But Maxwell's equations have persisted – they have survived its disappearance. Although the paradigm in which they appeared has foundered, they are still used, not only in numerous techniques, but also to construct the most up-to-date physical theories.

When an outdated physical theory is incorporated into another broader, theory, it is above all the argument that undergoes mutation. Its earlier version suddenly seems cursory and infantile: once the ether has vanished, the scientist can no longer understand what freak of fancy could have led him to conceive a substance with such ludicrous properties (it had to be extremely rigid, while offering not the slightest resistance to the rotation of the planets around the sun!) Yet the mathematical structure associated with it does not lose its value. The baby is not thrown out with the bath-water each time: scientists loathe waste.

Let us recall that at the origin of scientific revolutions we always find one or several paradoxes, which shatter certainties and open a range of questions. In extreme cases, as when removing one support from a house of cards, they precipitate a real conceptual collapse. Small causes, great effects: the mysteries of science can be volcanic. Quite often, it is paradox that shakes the tectonic plates of knowledge.

Because it prevents ideological glaciation, dynamises inertia, takes the axe to dogma, sends paradigms flying, upsets opinion and denounces prejudice, paradox has the power to change the law.

Truly, what is more healthy than a paradox?

11

The Orthodoxy of Paradox

> What is a paradox, if not a truth opposed to
> the prejudices of the vulgar, ignored by the
> bulk of mankind, and which current
> inexperience prevents their being aware of?
> What is a paradox for us today will be a
> demonstrable truth for posterity.
> *Diderot*

We have seen that the concept of paradigm is connected with that of consensus. It is the *doxa* of scientists, the highest common factor of their convictions. Aided by habit and success, every new theory gains in authority, becomes a *doxa*, and eventually becomes established as a very subdued version of the upheavals that installed it, even if it sometimes takes a while to circulate. After years of practice everyone becomes converted to the new crystallisation and the new science is transformed into a sequence of natural ideas. Gradually, it becomes acclimatised. *Doxa* becomes *orthodoxy*. If it grows too rigid, it may degenerate into *dogma*, and the paradigm turns into a machine for manufacturing new prejudices. As it degenerates, it deadens critical vigilance and wears down reservations. People get to like it, or at least they grow used to it. It fences off the closed field where prejudices will grow, and its power of conviction acts as a fertiliser.

It is only once prejudices have begun to flourish that new paradoxes can provoke a crisis of the paradigm.

Prejudices and paradoxes are sensitive to the passage of time. 'The paradoxes of today are the prejudices of yesterday,' wrote Marcel Proust in *Les Plaisirs et les jours*. Reciprocally, the prejudices of today already contain the paradoxes of tomorrow. Every prejudice is a potential paradox because paradoxes are defined by the prejudices they contradict, for instance during a confrontation with experience. To cut this story short, paradoxes are former prejudices, prejudices former paradoxes. In this context alternation is the name of the game.

The fate of paradoxes is a tragedy. In some ways, it is analogous to the fate of theories. Paradoxes have the fiendish and matricidal power to kill, at least in part, the theory that gave them birth. By a just reversal, only a new theory can in turn cause them to fade away. A theory arises to settle a paradox, and then dies of the paradoxes it kindles. The paradoxes born of new experience die of a new theory. One step brings them, another wipes them out. From offspring, they become corpses. The paradoxical state is therefore a temporary one. Ephemeral beings, transient interludes, paradoxes last only as long as it takes to transcend them.

But their brevity is their strength: it makes them *the fuel of scientific progress*.

12

Reactions to Paradox

> When one is right a day before the common
> run of people, one is taken to have had no
> common sense for a day.
> *Rivarol*

The identification of a paradox always promotes new
ideas, which are often greeted by gales of laughter
some of them aimed at those that will later prove
correct. This is because any new idea is likely to seem
bizarre or farcical at first sight. The human dimension
of science is never more evident than during the processes of transition induced by paradoxes. During
these turbulent periods, the objectivity of men of
science is not as cool as is supposed. There is no
shortage of examples. Luther, in his *Table Talk*,
described Copernicus as 'a madman who wants to turn
the whole art of astronomy upside-down'. The daring
hypothesis of Johannes Kepler, who linked the movements of the tides with those of the moon, also caused
great amusement. The idea that the tides were due to
the attraction of the moon could only be a joke.
Even the great Galileo guffawed, as if a cartoonist had
pictured a swollen moon busily sucking up the oceans.
We can also refer to a more recent affair. George
Uhlenbeck and Samuel Goudsmit are the two dis-
coverers of the spin of the electron who have gone

down in history. In reality, Ralph Kronig, a young American physicist of Hungarian origin, had the same idea independently and almost at the same time, in 1925. Unfortunately for him, he asked for Pauli's opinion on the subject. Pauli was already known for his achievements and insights, but this time he made a bad mistake and convinced Kronig that his spin hypothesis was ridiculous and unfounded. Impressed by the maestro's reaction, Kronig dropped the idea of publishing his article, unlike Uhlenbeck and Goudsmit, who had consulted no one. Soon after, it emerged that Pauli's objections were mistaken. Kronig should therefore have shared the credit for the discovery of spin. Richard Feynman, a born iconoclast, was right to define science as belief in the ignorance of experts!

Since new ideas make a noise that does not necessarily appeal to every ear, it sometimes takes courage to express the discordant ones. In 1597, Kepler sent Galileo a work favourable to the thesis of heliocentrism advanced by Copernicus. Galileo's timorous reply says a lot about the intellectual climate of the period. He wrote:

It was a long time ago that I turned towards the ideas of Copernicus. His theory allowed me to explain completely many phenomena which could not be explained by means of opposing theories, but to this day I have not dared to publish them for fear of suffering the same fate as Copernicus, who, while gaining immortal glory among the élite, was considered by most people as worthy to be hissed and mocked, so great is the number of fools. I might perhaps have dared to express my mediations if there had been more men like yourself, but as that is not the case I avoid approaching the subject.

Years later, encouraged by the friendship of the new Pope, Urban VIII, Galileo grew bolder and wrote his *Dialogue on the two chief systems of the world, that of Ptolemy and that of Copernicus*. In this work, he set out in simple terms his ideas on the Copernican system. To avoid presenting himself too overtly as Copernicus's advocate, he gave his book the form of a conversation between three people who discuss the pros and cons of the theory. However, he left little doubt as to his own convictions and his wit was sometimes caustic. The book appeared in 1632. A year later, Galileo, then in his seventieth year, was brought by his old friend the Pope before the Roman Inquisition. There, in a penitent's garb and on his knees, he was obliged to swear on the Bible that he abjured, cursed and detested the heretical idea of a fixed sun and a moving earth. He was put under house arrest, and his penance consisted of reciting once a week, for three years, the seven penitential psalms, which was no sinecure. With hindsight, therefore, the epistolary caution of Galileo was quite understandable, especially when one remembers that before him Democritus, Anaxagoras, Aristarchus of Samos, Giordano Bruno and many other innovators had been punished simply because their scientific concepts lacked the fireproof odour of sanctity.

This remark applies not only to a distant past. In 1920, an organisation was formed in Germany to oppose the theories of Einstein, who was attacked during a public meeting of this league at which he was present. Einstein was denounced as 'a publicity-seeker, a plagiarist, a charlatan and a scientific dadaist', in the thick of a poisonous atmosphere with swastikas and antisemitic slogans. On the other side of the Rhine, the physicist Bouasse called Einstein's imitators 'cock-chafers' and their works 'a buzzing of busybodies'. With more or less vehemence of tone, a great many other scientists (mainly those who did not understand it) rejected the new theory. Etiemble relates in his book *Eroticism and Love* that at much the same time it also took some courage for Soviet physicists to adopt the quantum theory when Stalin saw it as 'a Trojan horse of bourgeois individualism'. One may wonder how the quantum theory should have upset the little father of the people. According to Etiemble, it was because of Pauli's exclusion principle. This fundamental principle states that in an atom, no two particles of spin½ (two electrons) can exist in the same quantum state. In other words, because these particles lacked any gregarious instinct, they refused energetically to submit to the holy laws of collectivism. In Stalin's mind there could be no room for such individualistic particles.

Often, it is the scientists themselves who prove reluctant to accept new perspectives. They are cautious, and it must be admitted that history has taught them to be suspicious. It is normal and healthy for new ideas to be suspect to begin with, and for proof to be demanded of them, the more so since their first formulation is often obscure. The fierce debates they set going enable them to be stated more clearly, and their consequences inspected. In science, criticism is by nature constructive. But the reluctance of scientists may have deeper sources, which may be better sought in the sphere of private convictions. Scientists can be *philosophically* averse to the tenor of certain developments in their field. Like anybody else, they have their habits, fads and foibles, so it is understandable that they may harbour a nostalgia for the 'lost paradigm'. We have said already that Einsteinian relativity and quantum mechanics overturned our conceptions of space, time and matter. At the same time, they shattered the convictions of very many scientists, who felt as if the very foundations of physics were giving way, or in other words that they faced an intolerable threat to their livelihoods. As the ground seemed to be slipping from under them, it is understandable that some professional physicists succumbed to panic.

It is the probabilistic interpretation of quantum mechanics which gave rise, and still gives rise, to the heaviest turbulence. How could it be that God played dice? What really underlies the concept of wave function? But one also finds in the history of physics numerous cases of individual, often pathetic, hesitation. Michelson died without finding the condition that according to him would *rectify* his experiment on detecting the ether and confirming its existence. He was ready to accept that all the ether in the universe was pulled along by the earth, which meant granting a lot of prestige to one small planet. Rather than move the stool, Michelson was prepared to shift the piano. Earlier, Descartes, a logician if ever there was one, refused to believe that the fall of a body could be described by a simple mathematical law. In his view, this was an unacceptable return to the illusion according to which bodies obey intelligible natural tendencies. For him, only inertial motion was simple, and every other kind must be the average outcome of innumerable events of transmission of motion. This hits the anti-Newtonian jackpot.

One might easily extend this list of the sceptical, the lukewarm, the reluctant, and those disappointed by change, but that would be cruel and probably unjust, with too much help from hindsight. The more so since it is no doubt true that scientists have fewer prejudices than do other men, even though it may also be the case that they cling more tenaciously to those they have.

13

The Role of the Imagination, or Paradox Seen from Outside

Method, Method, what do you want of me?
You well know that I have eaten of the fruit
of the unconscious.
Jules Laforgue

As we have seen, in the course of its history science has proved again and again that it is capable, when necessary, of launching a process that engenders its own negation. Afterwards it reconstructs a new frame of thinking that reintegrates the old as it transcends it. But, contrary to a stubborn cliché that insists on seeing scientific creativity, especially in mathematics and theoretical physics, as the culmination of a purely logical approach, it appears in reality that this reconstruction rarely takes place in the innermost depths of the strictly scientific spirit. The automatic workings of science do not completely control its development. The fertile theory is not always a result that emerges from the linear logic of deduction. For example, one cannot reasonably persuade oneself that the Newtonian theory of gravitation emerged directly from the simple observation of a ripe fruit in free fall. Its unification of the laws of falling bodies (Galileo) and of planetary motion (Kepler) is not a straight spin-off from the apple falling in the corner of an orchard. It takes more than this altogether commonplace spectacle to unleash the idea that the moon and the fruit are subject to the same laws, that they are equally attracted by the earth, and to explain that, if the moon has a trajectory that never touches the earth, it is because it possesses a transverse velocity markedly greater than that of apples – a fact that is far from obvious to the naked eye.

Newton was certainly not the first man to have seen the moon in orbit and an apple fall, but he was the first to correlate their actions by showing that they are similar, which singles him out for special praise. 'One had to be Newton to perceive that the moon is falling on the earth, when everyone sees that it is not falling,' wrote Valéry in his notebooks.

There does not seem to be any logical connection between the facts and the theoretical idea that gives them meaning. 'Yes, but, damn it all, it's a certainty!' is an outburst, not a conclusion, and to be creative takes more than manhandling arguments and reasons. Discoveries are not made on an assembly line; invention is not merely dishing out *sinces* and *therefores*. Those who now believe that, in order to pioneer wave mechanics, it was enough to say: 'Since a quantum of light is associated with light waves, why not symmetrically associate a wave with the electron?' show a singular lack of perspective. Indeed, we must recall that when Louis de Broglie had this idea, he was still virtually alone, with Einstein, in believing in the light quantum, which was not yet called a photon (the idea of a quantum of light had not been taken seriously enough for physics to give it a name). So to imagine, as de Broglie did, that a particle could be diffracted deserves a comment of the same order as Valéry's on Newton.

It seems that to invent it is necessary to think, not literally, but beside, or in addition. A mine of scientific knowledge is certainly a necessary condition, but it is not enough, since new ideas do not germinate from the simple accumulation of ideas. A little shove is required, by which the moment of the *eureka* breaks free from its context. It is usually the imagination that takes care of giving the decisive flick.

Immanuel Kant foresaw this: before him, it was thought that imagination was a realm reserved for poets. 'This arrogant power, enemy of reason, mistress of error and falsehood,' as Blaise Pascal called it, could only deceive the scholar. This was the imagination run wild. The austere and fussy mind of the sage of Königsberg rehabilitated it into the sciences and raised it to the rank of governess. He discovered a poetic principle that lurked in the very act of forming concepts. A thinker who himself lived like an automaton, Kant turned this to account to reject the idea that the human mind was an automaton. For him, the mind was a free artist. This insight of Kant's has gained ground. It is accepted nowadays, according to the physicist Victor Weisskopf, that there is a 'Gödel's theorem' of science which recognises that it finds coherence somewhere apart from itself. As the creation of a new theory does not stem simply from listening to a rhapsody of facts or equations, we must recognise that it is a highly singular act of invention which requires more than an exercise of rote thinking in a formal framework. The movement of discovery goes from thought to reality, and not the other way round.

In the hours of great discoveries, a simple image may be the seed of a world. At decisive moments the scientist's mind functions by association of images, following a process that offers the fastest linkage system between the infinite forms of the possible and the impossible. One way or another, the pattern has grown familiar. Kepler, Newton, Cardano Kekule von Stradonitz, Einstein and many others have described the genesis of their best intuitions, always associated with a more or less controlled exertion of the imagination. In the case of Kepler or Newton, it even takes explicit detours via alchemy and astrology, a long way off the beaten track of the strict *cogito*.

This exercise of the imagination, sometimes close to conscious daydreaming, aroused in them a sudden intuition which detached them from ponderous stabilities and enabled them to see over the wall that blocked the view of their contemporaries. They found themselves looking on to the further side of paradoxes, to that zone of reconciliation where they cease to be paradoxes and take on new meaning. Only the imagination has this power of bringing to light a new way of thinking about facts already known.

The construction of a new theory is always a kind of uplifting that brings human thought into line with the hidden intelligence of natural laws. In Einstein's view, intuition sprang from a breath of the mind and was akin to divination. According to him, the quest for a faithful image of the world can only begin with an act of breaking free from data that fix and constrain us: 'Imagination is more important than knowledge.' Creation remains an act of provisional liberty, even when it concerns scientific activity. Here is Vassili Grossman, speaking in his novel *Life and Destiny* of his chief character, Victor Pavlovich Strum, a physicist by profession, who has produced a new theory:

> The theory had emerged, so it seemed, not so much from experiment as from Strum's head. The new had been born freely from this free play of intelligence, and it was this free play which had been somehow detached from experiment, and which had made it possible to find an explanation for all the wealth of old and new experiments. The experiment had been the external shock which had set the thought in motion. But it had not determined the actual content of the thought. It was staggering ...

It may seem paradoxical that logical rigour, which frames the scientific structure, should cease to function at the moment of its construction, or at least that it may become an obstacle. But it certainly becomes necessary again afterwards, in the rigorous and detailed staging of a demonstration, or else in the test of experiment, during which vagueness and fuzz is once again prohibited. But temporarily, during the phase of trying out the new keys, thought is less rigorously constrained; the rational dumps ballast, rigour dozes, dullness evaporates. For the time being, freedom becomes possible again.

Behind the equations hide sallies of the imagination, urgent perceptions that transcend logic and give science a measure of art. So there grows up a fruitful partnership between imagination and rationality. A game goes on between the questions and solutions produced by imagination on the one hand, and the constraints of coherent formulation and observation on the other.

Science is born from the tension between these two poles, but is not to be confused with either: it is neither pure constraint nor the systematic right to dream. It does not accept just any kind of theoretical innovation as cash in hand. The man of science, even if he is a devotee of new ideas, still holds that not all are correct, and there are screening systems to prevent the greeting of the new, its integration and reception, from becoming anarchic. But without the aid of the imagination, there would certainly be no science, since in order to progress it requires a stock pile of potentialities and hypotheses. There would certainly be professors, but not many inventors. It would only be possible to follow and apply the paradigms of the moment, or maybe insert a few new commas. But certainly not to transcend them.

Physicists must therefore combine the spirit both of invention and of method. It is the conjugation of these two antagonistic faculties that makes for great scholars: their theoretical innovations are all a delicately balanced stack of gravity and grace.

We have held forth at length and perhaps too long. It is time to get at the meat of the subject. Let us bring out our scalpels and prepare to dissect some of the amazing paradoxes of physics.

PART TWO

The Paradoxes of Physics

1

Wave-particle Duality, or the Metaphysics of Appearances

> Let us run to the wave, to spring forth
> again alive!
> *Paul Valéry*

In his fable called 'The bat and the two weasels', La Fontaine tells the story of a bat who unluckily falls into a weasel's nest. The weasel, who doesn't like mice, threatens to devour the bat. To defend herself, the latter explains that she is not a mouse since she has wings, which are peculiar to birds. Convinced by this argument, the weasel decides to let her go safe and sound. Two days later, the bat again falls into the nest of another weasel, this time one who dislikes birds. Threatened once again with being eaten, the bat explains that what defines a bird is not its wings but its plumage. Since she has no plumage, she must be a mouse, not a bird. In adapting her presentation to circumstances, the bat has twice saved her life. But when then is a bat, if it is neither bird nor mouse? The object of this chapter is to show that, in a sense, quantum particles have an ambiguity which resembles the bat's in La Fontaine. Their appearance depends on the context in which they are observed. What is called the 'wave-particle duality' is one of the most baffling aspects of the quantum world.

Light is readily associated with the concept of obviousness and simplicity, whose symbol it is. The book of Genesis affirms that the light was created on the first day, thus long preceding the ever-increasing complexity of the world. Yet this phenomenon, the most directly associated with the manifestations of the sensible world and with immediate visual perception, soon turns out to be very mysterious. 'We would know a great many things,' stated Louis de Broglie, 'if we knew what a light ray is.'

What is light made of? Isaac Newton, observing that the shadows of objects were sharp and not blurred, advanced the idea that light is made up of very tiny spheres, or corpuscles. He explained, for instance, the phenomenon of reflection in a mirror by the fact that the corpuscles rebounded off the mirror surface rather as a ball rebounds from a wall. He set out his views at length in his *Optics*, published in 1703. His contemporary, the Dutch physicist Christian Huygens, on the other hand, considered that light was a kind of wave, as he explained in his *Treatise on Light*, published in 1691.

If we stretch a cord tight and then shake one of its ends, we send out a wave that propagates along the cord although the cord does not leave our hand to run after the wave. Similarly, when a gust of wind sweeps over a cornfield, each ear of corn sways and oscillates at the mercy of the wind, but not one of them is torn off or carried from one end of the field to the other. Unlike corpuscles, waves carry nothing; they only transmit energy and information. Because there is a difference between the throwing of a stone and the motion of a wave, the wave and corpuscular theories seem absolutely irreconcilable. Nature must have chosen between the two. Is light a body, or else the motion of a body?

Newton's corpuscular theory initially supplanted the wave theory, thanks to the immense prestige of the man who had penetrated the secrets of gravitation. The simple idea according to which light consists of corpuscles that propagate in a straight line met a degree of success. Based on the concept of the light ray, it allowed an elegant explanation of all the phenomena associated with what is now called geometrical optics. But Newton's victory was not final. Right at the start of the nineteenth century the English physicist Thomas Young (who was also a doctor and Egyptologist) came out against it. He had observed that light has an odd way of interpreting the plus sign in arithmetic. Light superimposed on light may give darkness! This was only one instance of a group of phenomena known as luminous interference. Young published his findings in 1804. The corpuscular theory could not explain them, for how could one particle cancel out another? The wave theory did make them understandable: when two waves meet in water, there are certain points where the waves are always of opposite phase (one at its crest at the moment when the other is at its trough, and vice versa). Hence the two waves always cancel each other out at these points. If we now consider waves of light rather than water, the same effect results in these points being dark. Thus the wave theory contains the key to the mechanism by which light added to light can give darkness.

Young's notion was ridiculed at first. The scientific world erupted against him and his thesis. No one in England could understand how an Englishman could contradict the great Newton, and an article in the *Edinburgh Review* covered Young with insults. Yet after much resistance the wave theory carried the day. This spectacular reversal of opinion was in large part the consequence of the more complete studies undertaken by Augustin Fresnel after 1815. At the conclusion of his studies, Fresnel was able to propose a wave explanation for nearly all aspects of the behaviour of light.

In the second half of the nineteenth century James Clerk Maxwell showed, on the one hand, that all the known phenomena of electricity and magnetism could be understood on the basis of four equations (Maxwell's equations), and on the other hand that these equations admitted of solutions that corresponded to waves combining an electric and a magnetic field (and so described as 'electromagnetic') and propagating at the speed of light. Henceforth, it only remained to identify light as an electromagnetic wave, differing from other waves (such as radio waves) only because it had a much higher frequency. This perfect match between experiment and the theory of light made Maxwell's equations the second great mainstay of classical physics, on a level with Newton's mechanics.

Certain physicists then had the feeling that physics had achieved its aims, and that all that remained for the twentieth century would be to apply it, or to polish its calculations. But (happily for the twentieth century's physicists) at the end of the nineteenth century experiments revealed a number of phenomena which it could not account for (the most famous being the photoelectric effect). Because of these, Einstein in 1905 proposed a bold return to the corpuscular theory. He suggested that the interaction of an electromagnetic wave with matter takes place by way of indivisible elementary processes, where the radiation seems to be composed of corpuscles, or *photons*, each with an energy proportional to the frequency associated with it. Does this require abandoning the wave theory? It is tempting to answer Yes, since we have said that the two theories are mutually exclusive – if one wins, the other loses. In reality, the answer is No. Newton was right but Huygens was not wrong! This was the paradox, and it was to remain a source of great confusion until, in the 1920s, quantum theory rescued physics from its painful predicament.[1] To understand how much sheer paradox the idea of the photon contained, we must return to the extraordinary experiment of the two slits, which had provided Young with his most clinching arguments.

Imagine a machine shooting marbles at a wall pierced by two parallel slits set close to one another. Suppose that it shoots all its marbles at the same speed but in any direction. A little further away, boxes have been lined up to catch the marbles that pass through the wall. They serve to detect their final point of impact. Most of the marbles are stopped by the wall;

[1] Einstein was aware that this question would become the subject of an historic debate. In 1909, he said: 'I am convinced that the next phase in the development of physics will lead us to a theory of light capable of being interpreted as a sort of fusion of the wave and corpuscular theories' [quoted by Abraham Pias in *Subtle is the Lord*]. This remark, which went virtually unnoticed at the time, contains the premises of the quantum theory.

others pass through the first opening, either directly or by bouncing off one of its edges, while others do the same at the other opening. If, after a great many throws, we count the balls found in each box, we can determine how the probability of marbles' arriving varies according to the position of the box. The total number of marbles in a particular box is the sum of the number of marbles that have reached it through slit 1 and slit 2. If the first are shown in white and the second in black, the result obtained is that shown in Figure 4. In this case there is no interference.

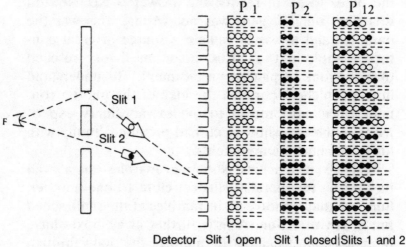

Figure 4. The experiment of the marbles and the two slits.
As each marble can pass through only one slit, those that pass through slit 1 can be shown in white and those through slit 2 in black. The probability P_{12} of a marble arriving in a given box when both slits are open is the sum of the probability P_1 that it arrives with only slit 1 open, plus the probability P_2 that it arrives with only slit 2 open. There is no interference.

If, as Einstein claimed, light consists of photons, i.e. little marbles, then a result should be obtained identical with what we have just stated, i.e. no interference. Now, as Young had seen, light actually does give rise to interference in this case. How is it possible that corpuscles should interfere like waves?

Let us repeat the experiment, but this time using electrons. The electron is an electrically charged particle (as its name indicates). It was discovered at the end of the nineteenth century by Joseph John Thomson, who photographed the trajectories of electrons travelling through a detection device called a Wilson cloud chamber. Going by the appearance of the tracks they leave in passing through a supersaturated vapour, there seems to be no doubt that the electrons are minuscule spheres which differ from marbles only in their tiny size. But back to our experiment. An electron gun fires electrons which all have the same energy in the direction of a plate pierced by two slits. The detector screen situated behind this plate is coated with a chemical agent that turns white at the point of arrival of each electron. What do we observe on the screen after firing a lot of electrons at it? One expects to find the same result as with the marbles, since one imagines that electrons are never anything more than small marbles. But surprise, the screen shows interference patterns analogous to those obtained with light (see Figure 5)!

The outcome is bafflement. On the one hand, Einstein tells us that light, which gives rise to interference characteristic of waves, is actually made up of photons. On the other, electrons, although they are supposed to be point-objects, also produce interferences! Are there little marbles in the waves and waves in the little marbles? Yet, as we have said, the two definitions – *the electron is a particle, the electron is a wave* – seems

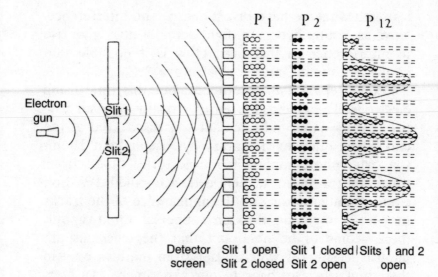

Figure 5. The two-slit experiment using electrons
If slit 2 is closed, the electrons are distributed in the detectors in the same way as the marbles. As it is certain that they have passed through slit 1, we can mark them white. If slit 1 is closed and slit 2 opened, they necessarily pass through slit 2, and can be marked black. But if both slits are opened, it is no longer possible to say through which slit each electron has passed (so they are marked in black and white), and an interference pattern is seen to appear.

mutually exclusive. Thomson wrote in 1925: 'The respective situations of the corpuscular and wave theories are those of a tiger and a shark. Each of these animals represents what is most powerful in its own element, whereas it is of absolutely no value in the element of the other.'

That electrons, photons and other quantum objects should behave both as waves and as particles offends common sense. What is anything if it is simultaneously both one thing and its 'opposite'? This is a genuine paradox, since it is impossible to visualise an object that is both at once. The concept of wave-particle duality is, to say the least, bizarre, but it will be immune to criticism if it proves to be coherent. We have to investigate whether the idea that an object might possess the appearance of a wave at one time and a particle at another does or does not lead to contradictions.

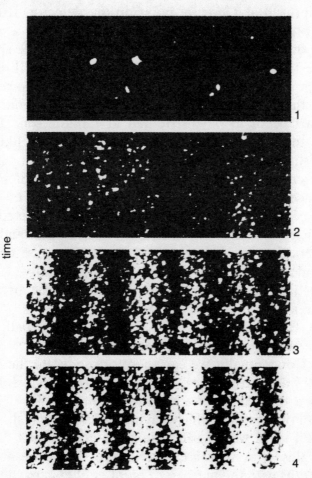

Figure 6. Result of the two-slits experiment conducted by firing electrons one at a time. The first electrons appear to have point impacts distributed in a random manner, then gradually we see the emergence of an interference pattern supposedly characteristic of waves!

Let us look at things more closely, and see what happens if we reduce the power of the electron gun so that its missiles emerge one by one and the emission and impact of each are isolated from those of the others. We now observe (Figure 6) that each electron is captured at a precise point on the detector. It does not break up into bits. So this is not purely a wave phenomenon, since a wave would fill the whole space. Note that in this experiment it is not known through which slit each electron passes, so we decide to represent them as half white and half black. The points of impact seem to be distributed at random, one here, one there, which makes the electron seem like an impish kind of corpuscle, a gambler. But, little by little, the screen records the growth of a system of interference fringes, in step with the accumulation of electron impacts! They are individually detected as particles, but it looks as if, between the gun and the screen, they are behaving like waves! Now we collect our ideas, if we have any clear ones remaining:

a) Since the electrons are dispatched one by one, these are independent phenomena.

b) Each electron, if it is really a little ball, has necessarily passed through one of the holes.

c) Conceptually, we can therefore separate the electrons into two batches: respectively, those which have passed through slit 1 and through slit 2.

d) For all those which have passed through slit 1, it is as if slit 2 was closed. The same goes for slit 2.

If this reasoning is correct, there should be at any point (as with the marbles) a number of electrons equal to the sum of those that have passed through slit 1 or slit 2. Now, as we have seen, the experiment shows nothing of the kind. Worse still, to open a second slit, i.e. to arrange a possible additional passage for the electron, is to prevent it from reaching certain spots (because dark fringes then appear). Which means that there is an error embedded in our mode of reasoning.

To clarify the situation we repeat the experiment, but this time we measure through which slit each electron passes, for example by illuminating it. We place a light source behind the plate which contains the slits, and arrange matters so that, when the electron passes through slit 1, a photon is detected deriving from 1, and the same for slit 2. With this arrangement we can 'colour' the electrons: white if they pass through the first slit, black if through the second.

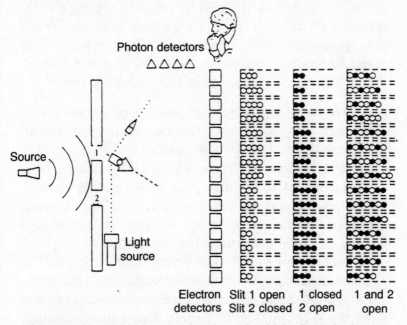

Figure 7. We decide to measure through which slit each electron passed, which allows them to be shown either as black or white according to the slit they pass through. But the interference figure has disappeared!

Let us first close slit 2. The (white) electrons are distributed as indicated in Figure 7, exactly as marbles would have been. We then close slit 1 and open slit 2. The (black) electrons are distributed symmetrically in relation to the white ones. We now open both slits. The result, once again, is disconcerting. We know through which slit each electron has passed, but the interference figures have disappeared! The behaviour of the electrons is identical with that of the marbles!

To sum up: if we observe through which slit the electrons pass, we no longer observe interference. It is quite another experiment. Reciprocally, if interference is seen, the question, which hole did the electron pass through? cannot be decided by physical means. Consequently, it was not valid to assume that one could simultaneously both observe interferences and measure through which slit each electron had passed. From this we can derive four basic observations:

1   We have to accept that, for an electron passing through one of the slits, the fact that the other slit is closed or not is crucially important. Somehow the electron seems to know whether the other slit is closed or not. The other path which was open to it, but which it did not take, exerted an influence on its point of impact. Remember how the mathematician Pierre Fermat described the behaviour of light: he explained that light which travels from one place to another follows the path which takes the least time. How does the light know in advance which path will be the quickest for it? Again, how can an electron know whether both slits are open or not, in so far as the word *know* has a meaning for a particle?

2   With no measurement to show which slit the

electrons passed through, they were capable of interference. After such measurement they were so no
longer. Somehow they were 'disturbed' by being
measured. *It is measurement that defined the system
measured.* Here we have a new property. In classical
physics we were accustomed to its always being possible to devise measuring apparatuses whose influence on the system is as weak as may be desired.

3 The notion of trajectory, which is so natural in
Newtonian physics, collapses before our eyes. In
observing interference, we cannot tell what path the
electrons took at each preceding moment. It also
seems that the concept of the location of a particle
has got to be discarded here.

4 For a particular electron, it is not certainly
known in advance at which spot it will hit the screen.
Now, the electrons are all emitted in the same conditions. So this destroys the classical idea according
to which the initial conditions completely determine
the subsequent movement of a particle. Classical
determinism is, if not dead, then at least badly
wounded.

Quantum formalism takes account of these trouble-
some facts by integrating the concept of *wave-particle
duality* proposed by Louis de Broglie in 1923. This
took further Einstein's idea that *there are corpuscles in
the waves*, by asserting the hypothesis that, *conversely,
waves are associated with the corpuscles*. All particles,
whether of light or of matter, show us a face that
depends on our way of looking at them. They cannot
be reduced merely to the aspect of their appearance.
It was from this finding that Niels Bohr, in 1927,

arrived at the idea that the two faces aspects of the wave-corpuscle duality are *complementary*.

This idea of complementarity is fundamental (and tricky); it accounts for the fact that the image of physical reality is not fully expressed either by the wave or by the particle image. In other words, it is necessary to realise that one is really dealing with two different but complementary languages to explain the phenomena of the quantum. In order to avoid all ambiguity, we must specify that, as used by Niels Bohr, the word *complementarity* is not to be taken in its usual sense. To elucidate the Bohrian significance of complementarity, we can go back to an idea put forward by the physicist John Bell (of whom we shall more to say in the chapter devoted to the EPR paradox). Imagine, with him, that we are photographing an elephant, first from in front, then in profile, and then from behind. One may then say that the different photos obtained are complementary in the usual sense of the word. They are consistent among themselves, and show us different aspects, partial but complementary, of the elephant (which may be considered as the synthesis or sum of all the images we can make of it). But, says John Bell, for Niels Bohr the word *complemetarity* has quite another meaning. For Bohr, complementarity comes close to the concept of contradiction, his conception being that, in analysing phenomena, we must use factors that contradict one another. If one establishes with certainty through which slit each electron has passed, this causes the interference figure to disappear. If we make no attempt to establish the course of each electron, the interference figure is left intact. Between these two extreme cases comes an entire range of intermediate cases in which the path of the electrons is established with a variable degree of precision. The more precise this measurement, the more the interference figure tends to blur; the more uncertain it is, the better the contrast of the interference fringes. Bohr's principle of complementarity may therefore be formulated as follows: The more we do so as to clearly observe the wave nature of light, the more we must abandon information on its particle properties.

Quantum formalism integrates this strange waveparticle duality into its very equations. But the trouble with that is that reality no longer presents itself to be seen as such. Any form of imagery seems to be barred. Who could just tell us what an electron looks like, fullface or profile?

2

Langevin's Paradox, or Journey Time

Travel shapes youth

Langevin's paradox, also known as the paradox of the twins, refers to the special theory of relativity constructed by Einstein during 1905. It had enormous repercussions at the time of its formulation by the French physicist Paul Langevin in 1911. Today, it undoubtedly remains the best-known of all the paradoxes discussed in this book. It is a classic of relativity, no doubt because it possesses all of the qualities a paradox needs to become famous. First, it is easy to describe and its content is both spectacular and disconcerting. Second, even though born in a strictly scientific context, it appeals to our imagination and transports us to a kind of science-fiction world in which time is no longer what we believed it to be. Finally, it is very instructive. It helps to understand the postulates and consequences of the theory of relativity, which are very far removed from what we are used to in daily experience. It is not by chance that Langevin's is one of the rare paradoxes to be taught in schools and universities, usually in the form of an exercise. Paradoxes and teaching always get on well together. Before claiming to have understood the theory of relativity, we must first examine all the situations where it has surprising implications.

Here is one such case. Consider two twin brothers, Rémi and Eloi, aged twenty. The first, Rémi, wild about space travel, decides to leave on board a rocket to explore part of the cosmos. He says farewell to his alter ego, and then takes off to travel in a straight line and at a constant speed to a planet seven light-years from earth. His rocket, which is faster than any known today, propels him at a respectable 296,794 kilometres per second, i.e. 99 per cent of the speed of light. As soon as he reaches his objective, he turns and follows the same course back again. His brother Eloi remains on earth, and is astonished to find that on his return Rémi is twelve years younger than himself! Rémi is only twenty-two, while Eloi is thirty-four. The two brothers are no longer twins; the sedentary one has become the elder. What can possibly have happened? This result, strange as it may seem, is perfectly in keeping with the theory of relativity, but we must admit that it requires some explanations.

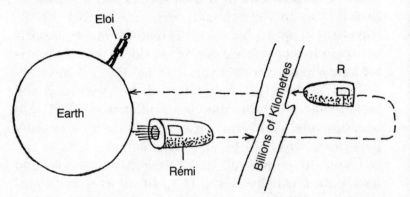

Figure 8. While Eloi stays at home, Rémi travels billions of kilometres in his rocket ship, turns round and comes home.

Everyday experience tells us that time is an absolute basic, that it is the same for all of us, flows at the same speed for everyone, and no one could possibly age more slowly than his neighbour simply because he made more journeys. The theory of relativity sees things otherwise. It postulates that the speed at which time passes depends on the speed of the location one happens to be in. How is it that we do not notice the relativity of time in everyday life? If it were true, surely it would be noticed? We would disagree about what the time is, and there would be confusion at railway stations before, during and after departure of each train.

In fact, Einstein's theory explains that, in order for differences between the time shown by two people's watches to be perceptible, their relative speed must be very much greater than any that we ever experience. So long as their relative speed is not an appreciable fraction of the velocity of light, both watches are synchronised. But in the case of the two twins relativity accounts perfectly well for the marked age difference noted on Rémi's return, because we have supposed that Rémi's journey is very long and that his rocket travels at close to light speed. If there is any science fiction in this paradox, it is only in this last assumption: we cannot make rockets that go as fast as Rémi's.

This astonishing prediction of relativity, which makes time an elastic quantity, has been confirmed in a spectacular manner thanks to the observation of particles called *muons*.

These are a sort of heavy and unstable electrons which are created in the upper atmosphere by cosmic radiation. When they are observed from the earth, they travel at a speed very close to that of light, but when observed in a laboratory where they are at rest, they decay so fast that it seems surprising that most of those that arrive near the summit of a mountain survive long enough to reach the earth's surface. In fact, however, the process of muon decay amounts to a measure of the passage of time; relativity tells us that their rate of decay is slowed by their speed. If the rate of decay is decreased, this means that their life-span is longer, and that they can therefore travel longer and further. Measurements have wonderfully verified these predictions.

What would be the viewpoint of an observer who travelled in step with muons? The muons being motionless in relation to him, he will not observe any decrease in their speed of decay. On the other hand, like the muons, he will find the mountain hurtling towards him at near-light speed. Worse still, the height of the mountain will seem to him less than it does to an observer watching from the earth, and he will thus have a shorter distance to travel to reach the ground, where he is in no hurry to arrive.

Now we return to Rémi and Eloi. Rémi, the travelling twin, is a sort of clock. Eloi, who stays at home, is another. If you find it demeaning to reduce human beings to clocks, imagine that each twin wears a wristwatch which indicates the passing years. These watches will confirm that they are not the same age. If Eloi could observe Rémi's watch while his twin is travelling, he would note that it is working more slowly than his own, and would have to conclude from this that Rémi is ageing more slowly than himself. So it seems that not only does travel shape youth, it also prolongs it! As it happens, it is a mistake to think that the travelling twin ages less quickly than his brother: in reality, both twins age at exactly the same rate. But how is it then that the travelling twin is younger than his brother when they meet?

It is possible to explain this strange outcome of relativity by means of an analogy. Two friends leave Paris for Dijon by car. One takes the most direct route, virtually a straight line. The other makes a detour via Nancy, taking a triangular course. Naturally, when they meet at Dijon the two friends find that their odometers do not show the same number of kilometres travelled, and this does not surprise them. We return again to Rémi and Eloi, and trace their trajectories, not in space but in space-time. We use only a single axis of space, the x axis which is the direction of flight of Remi's rocket (see Figure 9). The trajectories in this space-time start from the same point at the moment when Rémi takes off, and join again when the ship lands. Einstein invites us to imagine that the two twins' watches are the equivalent of the odometers of the two motorist friends. They show their wearers' 'proper time'. The stay-at-home twin follows the path AC. He is content to age without stirring, so that his path is a segment of the axis of time. As for the cosmonaut twin, he follows the path ABC, point B corresponding to the moment and the place where he turns back. Just as it was natural for the mileage of the two car-drivers to differ on their arrival at Dijon, so it is natural for the proper time AC to differ from the proper time ABC. And they do show different elapsed times, but there is a surprise. As path ABC is longer than path AC, we expect the travelling twin to be older than the other. In reality, measurement of the lengths of paths in space-time is made with a different 'ruler' than the one we use in our ordinary three-dimensional space. In space-time, ABC is shorter than AC, and Rémi returns younger than Eloi.

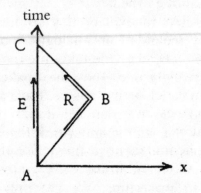

Figure 9.
E: Eloi's trajectory in the space-time linked to Earth.
R: Rémi's trajectory.

This prediction of the theory of relativity has been confirmed experimentally, but in a rather more general context, since gravity plays a part here. The principle of these experiments was as follows. An extremely precise atomic clock is placed in a supersonic aeroplane which is flown round the world. When it lands, the time the clock shows is compared with that on an identical 'twin' clock that has stayed on the ground. The latter indicates a flight duration very slightly greater than that given by its travelling twin (provided the aircraft has flown in the direction of the earth's rotation, otherwise there is a reverse effect). The difference between the two clocks is very small since, however supersonic the aircraft, it has a snail's pace compared with light. But the amount of this time difference corresponds very precisely to the predictions of relativity.

But the true paradox of the twins lies elsewhere.

True, the phenomenon we have just described affronts our intuition and upsets our conceptions of space and time. But after all, we need only to expand our intellectual outlook, and spend a few leisure hours reading up on relativity, to understand that this paradox contains nothing that is really problematic. The solution provided by relativity is a model of coherence. No, truly, the problem is not here. It arises rather from an objection that springs immediately to mind: the expansion of the durations, i.e. the slowing of the clocks, being a reciprocal phenomenon, each twin should observe that the other's watch is slower than his own. Let us put ourselves in the place of Rémi, the traveller. For him, it is Eloi who seems to travel. He sees him move away at first, then turn round, and finally come back. One may therefore follow the same reasoning as before, but this time taking Rémi's viewpoint, and conclude that when the twins meet again, it is the one who stayed at home who has aged by two years (because it is he whom Rémi saw travelling). In this case it is Rémi, on board his rocket, and not Eloi, who has aged fourteen years. The roles, and therefore the ages, are reversed. It was difficult enough to accept that the traveller might find his brother 'prematurely' aged, but relativity or no relativity, how can one accept that each of the twins is both strictly younger and strictly older than the other? Here we face a flagrant paradox that compels us to consider the two sides of the following alternative. Either we make the grievous admission that the theory of relativity contains a grave internal inconsistency, or else the reasoning we have just set out contains an error or a fallacy which had better be detected as soon as possible, since Eloi and Rémy are impatient to find out which of them has become the elder. In fact, even though it may bruise our ego, it is the second solution that we must adopt. Relativity emerges unscathed. It is we who made the mistake by assuming that Rémi's and Eloi's situations were identical. Actually there is a fundamental difference between the alternatives, which we will now examine.

Relativity, like classical mechanics, considers that there exists a privileged class of frames of reference, called Galilean Frames, in which all the laws of physics are expressed in the same manner. These particular frames are in translation rectilinear and uniform in relation to each other, which means that they move in a straight line and at a constant speed. To calculate the difference in age between Rémi and Eloi, we assumed that the reference frame associated with the earth is Galilean, which is a fair approximation. Then, when we reversed the roles by putting ourselves in Rémi's shoes, we implicitly assumed that the reference frame associated with his rocket is also Galilean. At first sight this hypothesis seems justified, since the rocket has a straight trajectory and a constant speed. But this is to forget that when he decides to return to earth Rémi must turn round. During this U-turn the reference frame associated with the rocket is no longer Galilean, since it is subjected first to a severe deceleration and then to a severe acceleration, whose values depend on the duration of the change of direction (if the U-turn is instantaneous, the rocket undergoes an infinite deceleration and acceleration). While he is performing this turn, Rémi is subjected to a braking force which may be colossal and which flattens him against the wall of his spacecraft. If, under these conditions, he could observe his brother Eloi back on earth, he would see that he does not undergo this deceleration force. That is the whole difference. Since Rémi is not in a matching situation to Eloi's it is an error to apply the same reasoning to him.

A more realistic calculation allows for what happens during Rémi's U-turn. We assume that he slows his ship progressively until it stops, makes his turn and then accelerates to resume his cruising speed. If the difference in age between Rémi and Eloi is calculated using Eloi's reference frame, which is Galilean, we may use special relativity. The result obtained will be a little weaker than if we assume that the turn is instantaneous (Rémi finds himself a little older then predicted). If we wish to redo the calculation in the non-Galilean reference frame of Rémi, it is necessary to enlist, not special, but general relativity, which makes all reference frames equivalent, whether Galilean or not. All we need do is to use a space-time metric that takes account of the accelerations. The age differences calculated from Eloi's and from Rémi's viewpoint are obtained in either case in the form of two integrals which are identical apart from a change of variables. Thus relativity emerges intact from this tale of the twins, as there is no inconsistency.

Travel slows the passage of time. But it does not bring it to a stop, unless it happens at the speed of light. The latter case applies only to particles of zero mass, such as photons, whose proper time does not elapse.

To the question posed by Lamartine in 'Le Lac':

Can we never on the ocean of the ages
Drop anchor for a single day?

relativity's answer is in the negative. But then again, poets are so demanding when they are in love!

3

Olbers' Paradox, or the Journey to the End of Night

Night comes, pirate night, disembarking on
golden skies.
Arthur Rimbaud

Some paradoxes have a greater spirit of resistance than others, and a kind of dignity. They demand a sustained effort and yield only to that. Olbers' paradox also known as the parodox 'of the dark night', is one of those stubborn enigmas which long plagued scientists. It was first stated in the eighteenth century, but defied repeated attempts at explanation for over two centuries. Only our most recent discoveries about the organisation of the cosmos have managed to shed a little light on the darkness of the sky at night. But why say that the dark night is a paradox at all? Isn't 'dark night' an obvious pleonasm, a redundant phrase? Surely it is normal for it to get dark at night?

From the beginning (say from the dawn of time), human kind has seen that in the evening, after sunset, the depths of the sky grow darker and darker between the twinkling stars. More prosaically, it is night. With the sun no longer visible, the celestial vault darkens. That is the most obvious of observations. Yet it poses a serious cosmological problem concerning the structure and evolution of the entire universe. For the simplest assumptions lead to the conclusion that the sky ought to be extremely bright even at night, and that we should be able to do without sunlight, but still see clearly for twenty-four hours out of the twenty-four.

Imagine that our universe is infinite, eternal, and uniformly populated with stars, and we are lying back to gaze up at a cloudless sky on a summer night. Then whichever way we look, our eyes must meet the surface of a more or less distant star. This can be illustrated by analogy with what a walker sees in a forest. If he stands in the middle of a very extensive forest, his line of sight is obstructed by trees in every direction. He cannot see anything outside the forest because his visual field is blanketed by trees. Of course, the further apart the trees stand from each other, the further the walker will see, and the further the forest must reach in order for its boundaries to be hidden from him. But however sparse the forest may be, there is a size above which it will form an opaque rampart. This also applies to the stars in our universe, for the cosmos is a forest of stars in three dimensions. The density of the stars is obviously much lower than that of the trees in a very dense forest, and that is why astronomers can see much further than walkers. But since we have assumed that the universe is infinite, they should nevertheless find a star in any line of sight, and the sky should appear to them as being completely paved with stars. That is the paradox, which conflicts with the most tautological of observations (it gets dark at night). The night sky is dark when it should be light.

One may wonder how it happens that so flagrant a

paradox should have gone unnoticed for so long. In particular, it is surprising to find not the least allusion to this paradox among the Greeks, though it is well known that they were not averse to questions relating to geometry. This calls for some explanation.

Broadly speaking, the Greeks left us three different philosophical systems, each of which offers a physical description of the cosmos:

– The *atomistic system*, derived from the Pythagorean and Ionian philosophers and taken up by the Epicureans, assumes that matter consists of fundamental indivisible elements, the atoms. The universe is described as being infinite, always identically repeated in space. The cosmos is a kind of unending litany of stars distributed at random. Overlooking lesser local variations, every region of space resembles every other region. Here we can recognise the premises of what is now called the cosmological principle.

– The *Aristotelian system* places the earth at the centre of the universe and makes the stars and planets revolve around it in perfectly circular orbits. The fixed stars mark the limit beyond which nothing exists – 'neither space, nor void, nor time', says Aristotle. The cosmological model is therefore finite.

– The *Stoic model*, developed in the fourth century BC, described the universe as a finite volume surrounded by an infinite void. This system cleared up the thorny problem posed by the Aristotelian system: how to conceive the absence of space, and especially the absence of a void, existing beyond a frontier which itself is defined spatially.

The Aristotelian universe reached its peak around 150 BC with Ptolemy. In his book *The Almagest*, he

summed up the knowledge acquired during the previous four centuries, filled in certain gaps in the Aristotelian system, and advanced a geometric vision of the
universe which was accepted virtually unreservedly for
1,500 years. The universe was geocentric, the earth was
fixed, spherical, and at the centre of everything, the
planets described small circles (the epicycles), with a
uniform movement, while the centre of the epicycles
was drawn in a circular orbit. Thus Western cosmology
remained largely Aristotelian until the fourteenth
century, before undergoing some adjustments. The fif-
teenth and sixteenth centuries saw the reappearance
of features derived from the Stoic and Epicurean concepts of the universe, but none of these different
models was capable of engendering the idea that the
darkness of the night sky was paradoxical. The paradox exists only if one accepts the hypothesis of an
infinite universe, eternal and uniformly populated with
stars. Consequently, the Aristotelians and Stoics would
have seen in the darkness of the heavens only a false
problem with a true solution: their own!

In the seventeenth and eighteenth centuries, science
went through a hectic and fertile period which culminated in the theory of gravitation. In this era there
were those who supported all three systems we have
mentioned. Kepler, for example, subscribed to the Aristotelian system. In 1610, he wondered about the secrets of the dark night, but soon concluded that it meant
that the universe was not infinite. According to him,
there were not enough stars to cover the far reaches
of the sky completely. Descartes took a more differentiated view. Rejecting the idea of a universe surrounded by a void (Descartes was allergic to the void),
he preferred the infinite universe of the Epicureans

but never mentioned the problem of cosmic darkness. As for Newton, he first believed in the Stoic system until his own theory of gravitation converted him to the Epicurean thesis. He 'needed' an infinite universe so that gravitation should not cause everything to condense into one great central mass. From that point on, the problem of the dark night could make its proper entrance.

The existence of the problem of cosmic darkness was explicitly addressed for the first time in 1721, by the English astronomer Edmund Halley (who gave his name to the most famous of comets). But he thought it could be instantly resolved by saying that most stars are so distant that their light cannot affect our senses. In his opinion, the eye could see nothing below a certain threshold of brilliance. This explanation will not stand up. Even if, beyond a certain distance, the stars are no longer detectable individually, the sum of the contributions of a great number of stars should be perceptible and the heavens should appear uniformly brilliant to us. It is precisely the accumulation of dim stars, invisible as such in themselves, which should produce brilliance in the night sky. Halley is not the discover of the paradox, but his reflections contributed to its emergence.

A little later, in 1744, a young Vaudois astronomer, Jean-Philippe Loys de Chéseaux, published a remarkable essay in Geneva. Starting from the hypothesis of an Epicurean type of universe in which all the stars resembled our sun, he calculated that the total luminosity of the celestial vault should be 90,000 times greater than the sun's! So the earth's temperature should be infernally hot, and we should see more than clearly at every hour of the day and night. Chéseaux then showed that the average distance of the stars which are not hidden by any material obstacle from the view of astronomers was of the order of 3 million billion light-years. This figure is so enormous that Chéseaux concluded that, however weak the absorption of the interstellar medium, it was strong enough to obscure the light of the remoter stars (he believed that space was filled with an absorbent fluid). This explanation was revived in 1823 by a German doctor devoted to astronomy, Heinrich Obers. He reformulated the paradox mentioned by Chéseaux in almost the same terms, but without quoting him. The paradox now bears his name, but is clearly misattributed.

For a century, it was believed that the absorption of starlight by the interstellar medium held the key to the paradox. But it became clear with the aid of the budding science of thermodynamics that this explanation would not wash, for in absorbing the radiation of the stars, the interstellar medium heats up until it reaches an equilibrium temperature at which it emits as much as it absorbs, and therefore does not reduce the luminosity of the sky. The supposedly defunct paradox was resurrected.

In 1848 an Englishman, John Herschel, proposed a different approach to the problem, which takes up Kant's speculations on the organisation of the universe and heralds the modern theory of fractals. It assumes that the universe is organised in a hierarchical manner; the stars are grouped in galaxies, with intergalactic distances far greater than the intragalactic distances. Then, the galaxies themselves form clusters of galaxies, separated among themselves by distances far greater than their individual size. These clusters are themselves grouped into super-clusters, the super-clusters are collected in clusters of super-clusters, and so on for an infinity of hierarchical levels. It can then be shown that, under certain conditions (in particular, there must not be too many stars in a cluster), such a universe is dark. Although it is infinite and contains an infinite number of stars, its sky is not completely carpeted with stars but only dotted with luminous points on a predominantly dark background. The paradox is thus solved in an elegant fashion, provided that we assume an infinity of nesting hierarchical levels, which is not easy to conceptualise. No doubt that is why the explanation met with no unanimous welcome. But already the fullness of the paradox of the night sky was glimpsed. Every science in succession tackled it and failed: optics with Halley, thermodynamics, and then geometry with Herschel. And that wasn't all.

At the end of the seventeenth century, measurements had already shown that the speed of light was finite. Astronomers knew that, as light travels only a finite distance in a finite time, observation of the stars reveals only their past, never their present. Thus certain stars which shine brilliantly in our sky have in fact been dead for many an age. From the earth, whenever we look at the sun, we are seeing it as it was eight minutes ago. That is the time it takes for light to travel the 150 million kilometres in between. Hence a quite trivial explanation of the paradox can be outlined. If the universe has existed for only a finite time, say T years, then the light of the stars situated more than T light-years away has not reached us. We are seeing only those stars that lie within a sphere of radius T light-years which is centred on the earth. If these stars are not numerous enough to cover the sky, it is natural for it to appear dark. It happens that we owe this very attractive explanation to none other than Edgar Allen Poe. No one before him had thought of it! At the start of this century, Lord Kelvin was to reformulate Poe's intuition in more rigorous fashion. Today, it is considered one of the main causes of the darkness of the night sky.

In the twentieth century, Einstein's general theory of relativity put an end to the classical concept of space by explaining that the matter and energy it contains give it a curvature. Material objects follow kinds of 'straight lines that bend', in response to the masses present in their vicinity. One might think that in this type of curved space-time the above-mentioned argument of the line of sight no longer holds. That is not so. The line of sight is itself also subject to the curvature of space. It can make the tour of the universe several times and end by stopping at the surface of a star.

Another solution proposed was that of the *red shift*, which is based on the model of an infinite universe, expanding, and of infinite age, in which the stars shine eternally, even if thermodynamics balks at this. The idea of expansion arose after Edwin Hubble had established with the aid of the Mount Wilson telescope that the spectrum of the galaxies is shifted systematically towards the red in proportion to their distance. Their light reaches us with a wavelength longer than at the moment of its emission. This comes from the fact that the universe is expanding. All its distance grow in the course of time. In particular, in the course of their intergalactic journey the wavelengths of the light of distant stars grow at the same rate as the universe. The more distant the star, the longer its light has travelled to reach us and the more its spectrum is shifted towards the red when it reaches earth. One may therefore imagine that, if the distance is great enough, the visible light emitted by the star reaches us with its spectrum so shifted as to be invisible to the eye. Although it may be carpeted with stars, the sky would then seem dark with the visual message of many stars lost in the invisible regions of the light spectrum.

This solution is very tempting, but, as if it were only fair for it to become its own victim ('everything shifts'), its status declined with the arrival of the Big Bang theory. The universe, described as the product of a gigantic cosmic explosion, is a kind of expanding sphere (at least for the time being) whose age is finite, of the order of about 10 billion years. In this new context, Poe's solution continues to apply. It was adopted by the American astrophysicist Edward Harrison, who was able to establish that the finite speed of light is the chief explanation of the nocturnal enigma, and that the red shift contributes only slightly to the dark of the night. Further, he showed, on the basis of an estimate of the total energy present in the universe, that even if they burned all their nuclear fuel, the stars could not shed a luminosity comparable to that of the sun, and that their lifespan is too brief, compared with the age of the universe, to saturate space with radiation. In order for that to happen, it would require the lifespan of stars and their density in space to be ten thousand times greater than they actually are.

So the astrophysicists are starting to see more clearly into the subject of the night. The next time we have the chance to look up at a starry summer night, we had better not forget to remind ourselves that if it is so beautiful it is because the universe had a beginning, and even the stars are transient.

Unless we prefer to consider simply, like Edmond Rostand in *Chantecler*, that 'it is at night-time that it is beautiful to believe in light . . .'

4

The Paradox of Schrödinger's Cat, or the Hazards of Boxing Clever*

> There is nothing to whip a cat for in the
> petty mischief it has just made.
> *D'Alembert, letter to Voltaire*

While describing Young's experiment with the slits (see part two, chapter 1) we saw that interference patterns only develop if we refrain from measuring through which slit the electrons travel. Those patterns result from the quantum superposition of the states *the electron passed through slit 1/the electron passed through slit 2*. Quantum mechanics describes this situation by representing the electron as a wave function whose structure allows for both possibilities. It is the sum of two terms, each of which represents one of the possibilities open to the electron. This superposition lasts only for as long as we make no measurements of the system, in other words as long as we do

---

*Another untranslatable pun: *les aléas de la mise en boîte*. Schrödinger postulated putting a cat in a box to see how its fate might be determined by quantum mechanics, thus creating the thorny problems described in the text. But *la mise en boîte*, literally 'putting (something) into a box', also means a leg-pull! (Tr. note.)

not seek to know through which slit the electron passed. But once we do make an observation, the superposition stops. A *collapse of the wave packet* happens, and the electron originally described as a wave that filled the whole of space appears on the form of a very localised impact. Observation compelled it to take a position. A priori, it was able to reach several different places on the screen, practically all except for the dark bands. The probability of its arriving at a particular spot could be calculated from the wave function. But of all the possibilities, only one is realised each time. The plurality of possible values is abolished by the actualisation of the electron's impact at a specific point on the detector screen: all but one of the possibilities are cancelled by the choice of a single one.

It is as if the measuring apparatus were a kind of lottery that draws one of the possible tickets out of the hat.

The quest facing us is to find out what is responsible for this collapse of the wave packet. What tips the transition from a sum of potential events to an actual result? What is it that precipitates the crossing from a formal rule (the principle of superposition) to a physical description of reality? Once having asked these questions, another puts its hand up: what did the electron's reality consist of, before it went through measurement? More generally, what is the concreteness of the reality that physics delineates?

These questions lie at the root of what is called 'the problem of measurement' in quantum mechanics. In 1952, Erwin Schrödinger, inventor of one of the fundamental equations in quantum theory, wrote: 'The problem of measurement remains the most delicate point, not to say the blind spot, of the theory, the one that cannot be filled by pure mathematics.' To this day it remains the cornerstone of quantum mechanics interpretations, each of which is characterised by the solution it offers to the problem of measurement, and by the conception of reality that subtends it. As yet, not one has secured unanimous approval.

As Schrödinger saw it, the collapse of the wave packet is like magic. He dismissed the very idea of a quantum jump. In 1935 he published an article in which he criticised the interpretation of the wave function in terms of spread of possibilities.

To illustrate his argument, Schrödinger invented a conceptual experiment, practically a cosmic joke, intended to demonstrate by means of a disconcerting scenario the paradoxical aspects of a naïve interpretation of the quantum formulation. From some angles this experiment resembles a game of Russian roulette, only with the choice of free participation removed.

Imagine with Schrödinger that an apparatus can detect the emission of a particle given off by a radioactive atom as it decays. It happens that the exact moment of the radioactive decay of an atom cannot be known in advance. In fact quantum mechanics describes it as a spontaneous process with no causal explanation, and provides no information about the exact moment when this decay will occur, no matter how much is known about the atom and its past history. But it does allow us to calculate the probability that the atom will decay in the following second or nanosecond.

Back to our subject: imagine also a box, and inside this box install a cat – in this case, *Schrödinger's cat*. Add to all this a mechanism so arranged that if the emission of the decay particle does take place, then it

Figure 10. The paradox of Schrödinger's cat.
The ambiguity of the state of Schrödinger's cat (after a drawing by Bryce de Witt in 'Quantum Mechanics and Reality', published in September 1970 in the American journal *Physics Today*).

causes a hammer to strike and shatter a phial containing a lethal gas. Straight away the gas fills the box and the cat expires with a last miaow. On the other hand, if the emission of the particle fails to take place (because the atom has not decayed), then the hammer stays untripped, the phial stays intact, and the cat stays alive. Having arranged these devices, we shut the box again.

The wave function of the complete system (box + cat + potentially lethal device) is obviously very complex, because this system contains a whole lot of particles. But after some time has elapsed, its wave function is the quantum superposition of the state *atom decays – hammer hits – phial breaks – cat dies* and the state *atom undecayed – hammer unmoved – phial unbroken – cat undead*. We have a situation analogous to that of the electrons in Young's double-slit experiment, where the interference pattern can be understood only if each electron is described as being the quantum superposition of an electron passing through slit 1 and an electron passing through slit 2.

In the context that concerns us, it is perfectly possible to calculate the probability that the atom is found decayed (and therefore the cat found dead) in the event of its being measured. The calculation is based on the wave function of the system, whose evolution in time is determined by Schrödinger's equation.

If we follow the quantum physics formalism, then so long as the result of the experiment has not been observed, the system is described by a wave function that superimposes the two states 'cat dead'/'cat alive'. In other words, the quantum concept describes the cat as being in an uncertain condition, neither dead nor alive, which an existential situation (let us face it) extremely difficult to conceive. Schrödinger concluded that quantum mechanics becomes absurd when applied to complex objects.

The question can be cleared up by saying that in any case this dead-and-alive condition cannot be observed 'as it stands', and therefore it is best to say nothing about it. 'Whereof one cannot speak, thereof one must be silent,' wrote Wittgenstein in the preface to his *Tractatus Logico-Philosophicus* in 1922. That is one point of view. We shall see that there are others.

The key point here is that according to the Copenhagen interpretation it makes no more sense to say that the cat is alive or dead than it does to enquire through which slit the electron passes in the slits experiment. Here again, we go astray each time we try to say too much about it. What we have to accept, or so it seems, is that the statement 'the cat is either dead or alive' is meaningless. In order to know conclusively what is the true existential condition of the cat, there is only one option: to open the box and take its pulse. The wave function then collapses: quantum superposition ceases. Of the two possibilities after the act of measurement (life or death), only one has come about, as if reality were not made flesh until observed. But how did the shift occur from the superposition of the two states to a single state, from a dead-and-alive cat to a cat that is dead *or* alive? Is it the fact of opening the box that determines the state in which the cat is found? Can the mere fact of observing it eject it from life into death? Can we keep a straight face while we try to imagine an eye that is sharp enough to kill so large an animal? Or was it already in the state in which we found it before the box was opened?

These questions may jar, or they may sound silly. Stephen Hawking says that when people talk about Schrödinger's cat, he reaches for his gun, which is quite a radical style of settling the question asked (nothing like a gunshot to jolt the cat out of its ambiguity), as well as being a very personal way of viewing the practice of theoretical physics. In reality, though, the question put by Schrödinger's cat is very delicate. Answers have been proposed that sometimes involve a particular interpretation of quantum mechanics, sometimes alter its formulation so as to render it – if possible – more 'acceptable'.

Here we shall sketch a quick anthology of these propositions, which makes no claim to be exhaustive.

For some physicists, the predicament created by this situation is so painful that the only solution consists of rejigging the quantum theory itself; from this angle it is only an approximate theory, unable to guarantee the coherence of its own rules, at least as regards the theory of measurement. These counter-propositions

exist in several versions. Because the quantum theory
has never been falsified by any experiment, they have
to meet a formidable challenge, namely to do at least
as well as a theory that works perfectly, but with different conceptual foundations – plainly no small
ambition.

Other physicists accept the quantum formalism as
it stands, but dispute the interpretation given by the
Copenhagen School, which they see as too minimalist.

Let us first examine a few of the counter-propositions. Under this heading come attempts that imply
a return to mechanicism, of which the most famous are
the *theories with hidden variables*. Several physicists,
among them Louis de Broglie and David Bohm, have
tried to construct them. There is a well-known joke by
the mathematician Hadamard that goes: 'Give me a
hundred parameters and I'll make you an elephant.
Give me the hundred-and-first, and I'll make it wag
its tail!' It is this idea, which is actually very serious,
that inspires the theories with hidden variables. They
try to resolve the problem by assuming that the final
state of the cat is governed by a parameter whose
value is unknown, but which pre-sets in advance the
exact moment when the radioactive atom decays. If
this moment comes before the box is opened, the cat
will be found dead. If after, then it lives.

These theories restore a fundamental determinism,
but they do so at the cost of difficulties that are not
to be minimised. They are often very complicated from
the mathematical viewpoint, and hence not very satisfactory aesthetically. On the other hand, the theories
called 'local' come up against a barrier demonstrated
by the physicist John Bell in the 1960s, and known by

the name of *Bell inequalities* (see part two, chapter 5, on the 'EPR paradox').

Another line consists of adding to Schrödinger's equation a little extra (non-linear) term which has the effect of making quantum superposition converge upon a proper, well-defined state when the dimensions of the system become macroscopic (on the microscopic level, the corrective terms are so tiny that they do not change the traditional quantum laws). This idea is relatively recent (1986), and is credited to G. Ghirardi, A. Rimini and T. Weber. It instantly gets rid of the problem of the cat, a macroscopic entity if ever there was one. More precisely, it prevents the problem from being put. But it does run into difficulties. For example, the differences it predicts compared with the usual quantum predictions have not been observed. So it is advisable to be more than sceptical towards this nonstandard model.

In reality, the frontier that divides those who accept the quantum formulation from those who do not is rather blurred. There is a border zone that contains those authors who invoke an effect of size to declare that the cat has complexity enough to reduce the wave function. More generally, measuring devices which are macroscopic are held to be capable of modifying microscopic states. It can also be argued that the cat is an 'object' big enough for any observer to be able to gauge its condition without therefore having to per-turb the animal. This amounts to saying that some-where between the objects of the atomic dimension and those of a detector's, quantum physics reverts to the classical rules. This interpretation poses the prob-lem of the whereabouts of this frontier between the microscopic and the macroscopic. Where is the customs post situated? At what level of complexity do we place the barrier that separates the beings or objects that do not reduce the wave packet from the beings or objects that can? Nothing in quantum theory provides a clear pointer, even though some suggestions have been made. The idea has been floated, for example, that there is not only a quantitative but also a qualitative difference between the macroscopic and the microscopic, connected to the fact that the equations of microphysics are time-reversible, whereas those of macrophysics are not.

In the same cohort of ideas, other physicists, such as Zeh, Zurek and Joos, bring in the environment (not in the ecological sense of the word), explaining that it continually measures the system in an unverifiable way. Since macroscopic objects have a very dense energy spectrum, with very adjacent levels, they are never completely isolated. In all rigour, it can never be said of a macroscopic system that it does not interact with its environment. In the particular case of the cat, it is this interaction with the environment that stops the dead-cat/living-cat quantum superposition. It acts as a very quick and efficient mechanism of disjunction.

We come now to those who, while accepting the formulation, reinterpret it in a different way from the Copenhagen School. Two interpretations, both of them very exotic, require to be cited. First, the thesis of the American Nobel prizewinner Eugene Wigner, who proposes a dualist solution. He considers that there are two realities, each one as ultimate as the other, namely the wave function (broadly speaking, matter), and consciousness (broadly speaking, mind). His argument consists in pointing out that everybody readily acknowledges that there is an effect of matter upon the mind, while assuming that there is no effect of mind upon matter. This, says Wigner, contradicts the general law according to which there is no action without reaction. He therefore floats the idea of an effect of consciousness on the wave function, which he sees as normal. By this account, it is the act of cognition by an observer gifted with consciousness that is responsible for the collapse of the wave packet: a transcendent act of consciousness intervenes to break off the superposition of the two states and impose the final state of the household pet, whose fate is somehow decided as the outcome of a 'mental operation'. This interpretation is underpinned by the idea that the human mind obeys laws of its own, different from those of ordinary quantum mechanics, which regulate only non-thinking matter.

The Wigner interpretation lacks a serious theory of consciousness, nor is there one in sight today. And there is another problem. Suppose that the box has been equipped with a camera which, starting from the moment t, takes a series of photos of the box's interior. Then an automatic scanning device analyses these photos, and if they are all identical (the cat has stopped moving) concludes that the cat is dead. A printer connected to the scanner then prints out 'the little cat is dead' on a slip of paper. The observer, who has so far remained passive, picks up the slip of paper without reading it, puts it in an envelope with his eyes closed, takes the envelope away with him and does not open it until a year later. It is not till this moment that he knows what has happened to his pet. There are some physicists for whom it is this act of awareness that triggers in the past the whole chain of events that we have just observed, and leads to the outcome described. Shall we say that the fate of the cat has remained suspended for all this time, or shall we conclude that the cat does not exist at all? To say the least, this explanation plainly contravenes received ideas. The fact that it is obliged to make the conscious mind a separate entity, not subject to the laws of physics, makes it hard to accept without further proof. It can be filed with other far-fetched speculations in the drawer which, in science, is never considered permanently locked.

Another thesis, equally exotic, not to say charming, has been worked out by a very restricted group of physicists who have contributed to the problem of measurement (and in this same context of the paradox of Schrödinger's cat) a solution even bolder than all the rest: the notion of *parallel universes*. This interpretation sounds like a hoax, but it is a hoax that has been taken seriously by serious physicists. Its first version dates from 1957, and we owe it to the American physicist Hugh Everett, who proposed it in order to banish a difficulty from quantum cosmology.

What does the quantum theory of parallel universes tell us? Simply that, when the measurement happens, there is no collapse of the wave packet to a single possibility, but a division of the entire *object measured + measuring apparatus* system into two distinct systems, in fact the creation of two parallel universes, one where the cat is dead and the other where it lives. At the moment of measurement the universe is duplicated to form two different branches that differ only in the condition of the cat, with no intervention of any awareness in the observer. So the measurement process no longer really exists, since all the possible outcomes are simultaneously realised. Because, in each branch, the result indicated by the measuring device can be read by observers, there is also a duplication of these observers, fortunately without their knowledge, since no communication is possible between the different universe-branches thus created. In particular, the observer in one of the two universes cannot put questions to the alter ego who has landed in the other universe. Moreover, it is for this reason that we would be unable to perceive these other universes. Duplication of persons, maybe, but no duplication of personalities, and hence no risk of schizophrenia. Each person knows only his own universe, in which he sees the play of the usual laws of quantum physics, including the collapse of the wave packet. As for the exact nature of the universe-branches, it is open to the most wide-ranging debate.

This protean theory (which comes in several variants) may seem far-fetched, but it cannot be trapped in the snare of its own logic. It is as hard to refute as it is to support. To its credit, we may count the fact that it is based on consistent mathematical foundations and that it eliminates the problem of the collapse of the wave packet without interfering with the quantum formalism. But for those with a fondness for parsimony, it has the drawback of leading to a fantastic procreation of universes, which stems from the immeasurable number of measurements of all kinds which have been made day by day since time immemorial. To invent new universes, they say, simply because the problem of measurement raises its head in our own is the utter antithesis of the method recommended by William of Ockham, armed with his razor for cutting back useless hypotheses. The thesis of parallel universes is burdened with metaphysical baggage that many consider too heavy. Few physicists refer to it today.

But the simple fact of its leading to theories as extraordinary as this one, vouched for by famous physicists, give the measure of the problem of measurement.

Out of this quantum cat-echism, we extract two lessons. The first is that it is unsafe to consider our customary classical world as a simple prolongation of the quantum world; hence the difficulty (and danger) involved today in describing the external world by relying on the fundamental notions of modern science. The second lesson is that the switch from a mathematical principle to a satisfactory description of phenomena is not reassuringly straightforward.

Jules Renard noted in his *Journal* that 'the ideal of calm is in a seated cat'. This is not valid for Schrödinger's, which just by lying quietly in its box managed to set off furious storms in the brains of physicists – not the usual 'ideal of calm'. To our knowledge, never has the health of a household pet made learned passions run so high, or caused so much scholarly ink to flow to the four corners of the world. In the fashion of Newton's apple, Schrödinger's cat has become a flesh-and-blood star of physics. Alive or dead? That is the question.

Schrödinger did not say whether or not, at the end of this conundrum, there was a prize to be won. But in this situation, even if the cat has got your tongue, there is no call to feel boxed in!

5

The EPR Paradox, or Romanticism on the Test-bench

> To be everywhere is to be nowhere.
> *Montaigne, Essays, I, 8*

Quantum physics raises the question of the accessibility of the real, and on this point the answers it suggests are very different from those of classical physics. Our aim in this chapter is to examine these differences in relation to a particular question, that of the locality (or separability) of the objects considered by physics. The locality of objects is an idea that seems self-evident. Every object appears to us situated in an exact location in space. When we say that two objects are situated in different places, we expect the effects of either object on the other to decrease with distance, and when we study one of them we are entitled to ignore the existence of the other. Each of them lives its own life, so to say.

This consequence of locality appears explicitly in classical physics: we have no need to pay attention to what is happening on Sirius in order to describe the fall of an apple in a terrestrial orchard, and we do not hesitate to consider that the acceleration undergone by an electron in an electrical field depends only on the value of the field at the point where it finds itself.

The absolute character of the idea of locality has been subverted by quantum physics, whose basic tool is the *wave function*. Now the wave function of a particle at any given moment is generally 'spread out' in space – hence the idea, first put forward by Schrödinger, that the quantum formulation is *non-local*. Since then, this non-locality, which clashes head-on with common sense, has been established by experiments.

This finding is the unexpected fruit of a gripping controversy that pitted together two giants of this century's theoretical physics, Albert Einstein and Niels Bohr. It was unexpected because at first their arguments did not cover the specific question of locality, but involved the completeness or otherwise of the quantum formalism. The problem lay in knowing whether or not reality included elements that were not contained in the formalism. Is it imaginable that the wave function of an electron does not contain everything that can be known about it?

Even though he was a pioneer in the understanding of quantum phenomena, Albert Einstein never accepted the current interpretation of quantum physics, which is known as the *Copenhagen interpretation*. It seemed to him to be incompatible with the idea of an objective reality, independent of the knowledge obtainable about it. Furthermore, the father of the theory of relativity could not bring himself to accept that quantum theory was incapable of predicting anything but probabilities. In his view this powerlessness laid bare the partial and roundabout character of the quantum formulation. Only a few years before his death, he was still writing to Louis de Broglie: 'I must resemble an ostrich that keeps on hiding its head in the sand so as not to face up to the nasty quanta.'

It was on these questions that he differed with Niels Bohr, in a titanic but friendly debate whose terms may be summed up by the following two questions:

1 Does quantum mechanics provide complete information about the *phenomena* that it claims to describe?
2 Does it give a complete description of the world?

Bohr rated these two question as synonymous (phenomena *are* the world), and answered both with a categorical Yes. Einstein conceded, if pushed, an affirmative reply to the first question, but refused to believe that it could contain a definitive account of reality. He saw quantum physics as an ingenious theory, acknowledged its operational efficacy, took note of its practical scope, but denied that it could depict the inward structure of reality as it exists apart from ourselves. It was usable only as a makeshift expedient. Einstein wrote to his friend Max Born: 'I believe that the present formulation is as true as, for example, the statements of thermodynamics.' In fact thermodynamics (non-statistical) predicts the properties of matter in equilibrium, but is incapable of describing phenomena such as thermal fluctuations, Brownian motion, and various other things that can only be explained by allowing for the atomic structure of matter. In one sense, it is merely the art of finding averages.

One motivation for Einstein's allergic response to quantum mechanics was the theory's non-deterministic character. According to him, a *good* physical theory must not only seek to combine laws and empirical observations into one and the same account (which quantum mechanics actually does very well). He also required it to have an internal perfection. Ideal theories in the Einsteinian sense should eliminate chance, if not from all their structures, then at any rate from their principles. Now quantum theory stipulates that, from knowing the initial state of a physical system, it is not generally possible to predict with certainty the results of a measurement performed upon it, but only to calculate the probabilities of obtaining a particular result.

A second reason for Einstein's opposition derived from his attachment to the ordinary realism of physicists. He wanted to regain an objectivity less ambiguous than that of quantum physics. For him, it was a matter of rediscovering (or reconstructing) the idea of a real world whose tiniest fragments exist objectively, in the same sense that tables and chairs exist, whether or not we happen to be observing them.

In addition, Einstein adhered firmly to the principle of 'local causality', which decrees that distant events cannot exert influence on nearby objects without a mediating agent of some order or other. This conviction was shared by the majority of physicists. To gather what it means, we will take the example of a pair of particles. Einstein considers that such a system is necessarily *separable*, i.e. that it is permissible to talk separately about each of the particles that compose it. This amounts to accepting that it is possible through thought to divide any system of two particles into two subsystems (each of the particles) each of which constitutes a totally separate physical entity, in the same way as the overall system.

As for Niels Bohr, he rejected this hypothesis of separability, with arguments that could seem hazy at the time, although his intuition turned out to be correct. In any case, he regarded the probabilistic character of quantum mechanics as basic and irreducible, which in his view ruled out the idea that particles were 'things' with well defined properties under all circumstances.

Plainly the debate in question deals with fundamental issues. The genius of these two men, Einstein and Bohr, was to give it all the scope that it deserved. Could it be, as Einstein thought, that quantum mechanics was a theory valid for describing the average properties of matter but incapable of describing the reasons for the behaviour of individual systems? Or must it be agreed, in line with Bohr, that through the medium of the wave function quantum physics tells us everything that is knowable about physical phenomena? Quantum mechanics has grappled with these questions from the moment of its birth.

Einstein's objections concerning the completeness of the quantum formalism culminated in 1935 in an article published in the *Physical Review*, which expresses the essence of what has since been called the EPR paradox, from the names of Einstein and his two colleagues of the moment, Podolsky and Rosen. This seminal article has without doubt been the oftenest quoted and most debated in the entire history of physics. Enlisting a favourite method of his, Einstein attempted to highlight the incomplete character of quantum physics with the help of a 'thought experiment'. Without raising the slightest hypothesis about determinacy or indeterminacy, the thesis he framed was intended to show that quantum theory does not tell us everything that we have the right to expect in a *good* physical theory.

Einstein's initial hypotheses are:

a)   The predictions of quantum mechanics are correct.

b)   No influence can propagate faster than light.

c)   Lastly, Einstein adopts a very broad criterion of objective reality, which he formulates as follows: 'If, while in no way disturbing a system, it is possible to predict with certainty (i.e. with a probability equal to unity) the value of a physical quantity, then there exists an element of physical reality corresponding to that quantity.'

For Einstein, there are therefore facts, events, that are real; they are not just 'realities for us', but genuine 'realities in themselves', which can be distinguished from appearances by their predictability.

Einstein goes on to explain that for every element of physical reality (e.g. the speed of a particle) there must be a corresponding magnitude defined by the formalism, whether this magnitude is measured or not. It is on this sole condition that a physical theory is called 'complete'.

The application of the criterion of physical reality to the quantum theory leads to what has been called the EPR 'paradox', which really counts as a genuine paradox only for those who expect from quantum mechanics a complete image of the atomic world. In fact, what Einstein shows is that the combination of the three hypotheses *a*, *b* and *c* applied to a certain thought experiment leads to attributing to the sub-systems properties not allowed for by the quantum formalism. From this he deduces that there must exist a sharper level of focusing on physical reality, which remains to be discovered.

According to the Copenhagen interpretation, two so-called 'conjugated' quantities, such as position and momentum, have no simultaneous reality owing to the physical interaction created by the act of measurement: since measuring position modifies the state of the system, its momentum is not measurable simultaneously. This finding flows from Heisenberg's indeterminacy relations, more precisely from the 'noncommutation' of the quantum operators associated with position and momentum. If the momentum of a particle is the quantity taken as corresponding to an element of physical reality, then the coordinate of position cannot be ascertained with certainty and hence, by the criterion defined above, does not correspond to a physical reality.

Einstein rejects this conclusion. Granting his hypotheses, the experiment he imagines shows that there is a means of knowing both position and momentum. It follows that position and momentum must each correspond to an element of physical reality. As the formulation of quantum mechanics does not take both of these two elements of physical reality into account, then it is incomplete. Bohr's reply to the article by Einstein, Podolsky and Rosen was published by the *Physical Review* in the course of the same year, 1935. Like most of his articles, it is very hard to read. To simplify, it may be said that the solution it offers to the EPR paradox is contained in the idea that hypothesis *b* (above) is not totally acceptable, and that Einstein's criterion of reality is unsound. Bohr explains that it is impossible to achieve a really sharp separation between the behaviour of atomic objects and their interaction with the measuring devices that define their conditions of existence. This means that the velocity of a particle, for example, is not a property of the particle but a property it shares with the measuring instrument. From this Bohr deduces that it is necessary to abstain altogether from reasoning about unobserved objective reality.

It is obvious what differentiates Einstein's view from Bohr's. For Bohr, the quantities considered have no simultaneous reality. For Einstein, they do have one, but quantum theory is not complete because it fails to provide an account of it. This Bohr-Einstein controversy has a philosophical status because it deals with our conception of the world, the human idea of it, and the role of physical theories. But it also belongs to the sphere of physics, in the terms formulated by Einstein, because he maintained that what he saw as the incomplete nature of quantum mechanics implied the eventual development in future of a 'better' physical theory, i.e. one that gave more information. This is a question that experiment has been able to settle, and in a way unfavourable to Einstein's hopes. It was elucidated in two stages: first by a theoretical discovery made in 1965 by the physicist John Bell, and then by several corroborating experiments, the first not very conclusive, but almost irresistible in the case of the three experiments conducted in 1983 at the Institut d'Optique d'Orsay by Alain Aspect and his team.

Unfortunately from the human angle of the affair, by the time of the discovery of Bell's theorem, both Einstein and Bohr were dead. What would they have said about it? We shall never know.

We have said that Einstein's three hypotheses implied the incompleteness of quantum mechanics. That means that if these hypotheses are correct, then quantum mechanics must be either replaced or completed. But after all, are these hypotheses correct? Yes, said Einstein: good sense and the particulars of physics, especially of relativity, combine to prove that they are.

The physicist John Bell was a very staunch advocate of an explicitly realistic cosmology. For this reason he was inclined at first sight to side with Einstein. Yet he also noticed something very curious, which was that the theory of Louis de Broglie and David Bohm, whose purpose is to complete this mechanics through introducing supplementary (also called 'hidden') parameters, manages to satisfy hypotheses *a* and *c* only by violating hypothesis *b*. The question of non-locality arose. Was it a flaw in this particular theory? For someone holding Bell's ideas, this was the most attractive conjecture. But Bell was aware that the ideas that look most attractive at first sight are not always necessarily true. So he set out to study the question in a broader perspective.

He studied it so well that to the surprise of many, and his own in particular, he was able to show that the conjecture was false. Not only the theory of Louis de Broglie and David Bohm, but likewise any theory purporting to describe reality and satisfying hypotheses *a* and *c*, must of necessity violate hypothesis *b* or a hypothesis of very similar scope. That is the content of Bell's theorem. It is proved by showing that any realist theory satisfying hypotheses *a, b* and *c* involves restrictions related to the results predicted for certain measurements of correlations, restrictions now termed Bell inequalities. To give an example of what such a restriction may be, the Bell inequalities specify for example that a census of the French population will never be able to arrive at a number of black people who can ride a bicycle greater than the number of black men added to the total number of women cyclists present in that population. In other words, John Bell called attention to particular situations, called EPR situations, for which the predictions of quantum mechanics came into conflict with Bell's inequalities. This is the case for phenomena said to be in *strict* correlation.

For two phenomena in strict correlation, a theory that satisfies hypotheses *a, b*, and *c* necessarily contains hidden variables, and would yield definite predictions if the values of all of these variables were known. As they are not known, it can only offer probabilities. Several of these can be defined: that of obtaining a particular result for the first phenomenon and another for the second; also that of obtaining a given outcome for the first result when one takes no account of the second, and so on. Bell correlated these various probabilities to construct a simple mathematical magnitude, about which he demonstrated that it is necessarily less than 2 if the consequences of locality are accounted for. The quantum mechanics of the Bohr persuasion predicts in these circumstances a value greater than 2, in violation of the Bell inequalities, as if there were a greater 'conspiracy' or 'cooperation' between the correlated particles.

An experimental proof became available, which would either confute quantum mechanics and its description of a quasi phantasmal reality, or else rule out the existence, somewhere in its shadow, of a theory that would endorse a reality *locally* defined.

The experiment ruled firmly in favour of quantum mechanics.

The test proposed by Alain Aspect in 1976 called for studying the correlations between photon polarisations. He focused on measuring the polarisation of the light emitted under certain conditions by calcium atoms. The term used is 'radiative cascade': two successive photons are emitted by the atom, and their polarisations are strongly correlated because they derive from the same atom. Simple devices (polarisers) exist that will let light through if it has a certain polarisation, and block it if it has another. By counting, for each type of photon, the numbers that do and do not pass through, we evaluate the various probabilities mentioned above. If the statistics are sufficient, we have a laboratory method of testing Bell's inequalities. The findings of the experiments are unambiguous. The Orsay team used the count-rates recorded to calculate the value of the mathematical function that, in order to satisfy Bell's inequalities, must always be less than 2, whereas quantum mechanics predicts a value less than 2.7. The experiment yielded 2.697, in perfect agreement (allowing for residual uncertainties) with quantum mechanics. Bell's inequalities are therefore violated in this case, and violated in accordance with the predictions of quantum mechanics. Hence we must say goodbye to interpreting quantum mechanics in line with Einstein's ideas, alluring as they are, and admit that the world conforms closely to the predictions of quantum mechanics, even when those predictions seem extraordinary. Yet again, physics gives a slap in the face to common sense, which must be starting to get used to it.

The astonishing character of this finding emerges if we return to the premises of the EPR argument. In fact, the conclusion of the experiments, allowing for Bell's theorem, is that hypothesis *a* remains valid, but on the other hand we have to abandon at least one of hypotheses *b* and *c*, with their relativist inspiration.

According to relativity, no signal can be propagated faster than light. Therefore, thought Einstein, there are cases in which we may be utterly certain that neither of two events can influence the other. These are the cases where the two events are so distant in space, and so close in time, that light has no time to connect them. It is this hypothesis that is called Einstein's principle of separability. In this sense it may be said that Aspect's experiments established the existence of non-separable systems, as predicted by quantum mechanics. We have to accept the (very romantic) idea that two photons that have interacted in the past compose an inseparable (or 'entangled') whole, even when they are very far apart! They have an overall behaviour that cancels out any prospect of explanations made in terms of individualist photons each possessing in itself, and in advance, the property that one has decided to measure.

What is this non-separability made of? Is it based on a mechanism? Does it revolutionise our conceptions of space and time? The least we can say is that it is unexpressible by a simple image, so far does it transcend our familiar concepts. And yet we must grow used to living with it, and be prepared to think it.

6

The Violation of Parity, or Faint Praise for Difference

> Tell this story to an old stick, and it would
> regrow leaves and roots.
> *Henri Michaux*

Several hundred different particles are now known, and we tend to anticipate that a world so rich will be formidably complex and difficult to grasp. In reality, the study of the decay of these particles has shown that regularities and associations abound. Their behaviour is not as unpredictable as might be expected. The magical concept that permits this simplification is that of symmetry.

Symmetry is familiar in art and architecture, and is also found in nature, for instance in the shape of a snowflake or the spherical form of the sun. Asymmetry is also present in the world around us. We know for example that the DNA molecule, which contains the genetic code of living beings, always has the shape of a right-handed helix. There is no such thing as a left-handed DNA molecule. DNA is said to *break* left-right symmetry. Our heart, which we usually wear on the left, does the same.

The existence of symmetrical structures reflects underlying asymmetries in the laws of physics. For example, physicists long ago discovered a close link between the geometric symmetries of space and the laws of mechanics. The symmetry of translation of space, which states that physics is the same whether you are in Saclay or Geneva, leads directly to the conservation of the momentum of particles (this is the principle of inertia, which prohibits spontaneous modifications of movement), while the symmetry of rotation implies the conservation of kinetic momentum. The symmetry of translation of time, which stipulates that one moment is as good as another, and that physics will be the same tomorrow as it is today, leads to the conservation of energy. Thus, the most fundamental laws of physics flow from the elementary fact that space is void and uniform. This link between the symmetries of space and time and the dynamic behaviour of material objects has the beauty of a miracle.

But the converse is false: an observed asymmetry does not necessarily imply asymmetric laws. A left-handed DNA molecule would be just as compatible with the laws of physics. There really are people with their hearts on the right. But let us take an example based on a simple experiment. Stand a bar on a table in a vertical position. This situation is symmetrical in relation to the axis of the bar, since it can be made to rotate around this axis without changing its physical position. Now let go of the bar. It falls on the table in a particular direction, and in doing so it breaks the symmetry of rotation of its initial position.

Particle physicists have found themselves introducing symmetries which have no analogue in daily life. These abstract symmetries play a fundamental part in the organisation of elementary processes, introducing a little order and discipline into what, without them, would be a gigantic chaos. In addition, they provide the key to recent ideas on the organisation of the forces of nature. Some say that this rise to power of the concept of symmetry in modern physics is the revenge of Plato over Democritus – the victory of the *idea* (symmetry) over the *thing* (the particle).

All particles carry a certain number of *labels*, which class them in different categories and are closely linked to the way they behave. Some labels, like mass or electric charge, are only miniature versions of properties familiar in the macroscopic world. We all have a mass and a weight. Other labels, like baryon number, lepton number, parity, or still more 'strangeness', seem to us much more mysterious because they are not manifested on our scale. We do not manage to read these labels with our macroscopic apparatus because they do not produce detectable forces. But the particles can read them, and respond appropriately.

The labels enable the particles to recognise each other and to react 'accordingly'. All particles interact with one another with more or less force, from the barely caressing brush of the neutrino to the explosive annihilation of the proton and antiproton. Perception of the labels varies according to the 'sensorial' equipment of the particles. Not all see them all – their view is selective. Neutrinos, for example, do not distinguish electric charge but they do see the lepton number. Protons see both electric charge and baryon number.

A question arose for those who wanted to understand the organisation of the four fundamental interactions of nature: gravitational force, the weak force, electromagnetism and the strong force. Do all these forces respect the geometrical symmetries of space and time? The issue is clear for electromagnetism and gravity. Maxwell's electromagnetic theory explicitly incorporates all the symmetries previously discussed, as does Einstein's theory of gravitation. Physicists have long assumed that the nuclear forces also respected all the geometric symmetries. It would have been very strange if the laws of the conservation of energy, of linear momentum and angular momentum, were violated in the microscopic world. Anyway, the facts have confirmed this expectation. From this viewpoint, particles are nothing out of the ordinary. They behave like everything else. All the events in which they take part conserve energy, impetus and angular momentum. But there are other geometric symmetries which belong to a different category. These are space inversion and time reversal. These symmetries are *discrete* and not continuous, and therefore are less familiar to us than the previous ones. But physicists believed them to be just as well verified by the four fundamental interactions.

The law of conservation which states that the universe is invariant by any reflection in a mirror, i.e. that nature displays no preference for right or left in physical laws, is called the *conservation of parity*. It signifies that the image of any experiment in physics seen in a mirror is also an experiment in physics, and it states that the laws of the world are not changed when seen through the looking-glass.

In the 1920s, it was discovered that the wave functions of atomic electrons have a well defined parity, positive or negative, i.e. that they are either odd or even in relation to the variables of space. It was also known that the parity of a stationary quantum state does not change with the passage of time: a wave function cannot change from positive to negative parity. Parity is said to be conserved. These ideas were generalised to the physics of particles, by attributing to each particle an intrinsic parity which determines the properties of reflection of the system which creates or destroys it. In 1927, Eugene Wigner showed that the conservation of parity simply indicates that all the forces between particles are without any right-left tendency. In other words, any infraction of the conservation of parity would be equivalent to a violation of right-left symmetry. For the reasons mentioned above, nature seems to be completely ambidextrous. It is not lateralised. The asymmetric systems encountered therein are not evidence of any fundamental asymmetry. The fact that our heart is situated on the left is an accident of the evolution of life on our planet, and does not imply any asymmetry of natural laws. In theory, it could very well have happened that we had our heart on the right.

The law of the conservation of parity can be illustrated as follows. Imagine that we have been filming an experiment in physics. After the recording, we run the film backwards and project it on to a screen, which now shows a reversed image of what really happened. To claim that there is conservation of parity simply comes down to saying that one cannot tell, when watching this projection, whether the film has been reversed or not. In other words, invariance by the operation of parity (designated P) is satisfied if object and image have the same probability of being observed in the real world. In other words, if physics respects this invariance, it is because it cannot tell right from left.

At least, that is what was believed until 1956. In

that year, physicists began to rack their brains over the *theta-tau puzzle*, which involved the weak nuclear interaction (the weak force is manifested during certain particle decays, such as beta decay, during which radioactive nuclei emit electrons or positrons). In 1947, two English physicists, G. D. Rochester and C. C. Butler, had discovered in cosmic radiation a new neutral and unstable particle, the K^0 meson. A year later, another physicist, C. Powell, discovered a charged partner now called the K^+ meson. These two discoveries marked the beginning of an eventful story, that of the *strange* particles. First, the positive K meson seemed to appear in two different forms. One, the θ (theta) meson, decays into two π mesons. The other, the τ (tau) meson, gives off three π mesons. Except in this single detail, there is no way of distinguishing the θ from the τ. They have the same mass, the same electric charge, and the same lifetime. It would have been very tempting to say that there was only one, capricious, K^+ meson, which sometimes decayed into two π mesons and sometimes into three. The problem arose from the fact that the θ meson has an even and the π meson an odd parity. As we know that two π mesons have a global even parity, then parity is conserved in the decay of the theta meson. But the global parity of three π mesons is odd, and therefore different from that of the θ meson. Hence, the τ meson and the θ meson cannot be one and the same entity. This led to the acceptance that the two versions of the K meson were really two distinct particles, but ones impossible to tell apart except by their parity. It might also have been concluded that parity was not conserved in one of the two decays (that of the θ meson), but this hypothesis was unimaginable for most physicists in

1956 because it amounted to admitting a violation of right-left symmetry. The conservation of parity had been established for all the other interactions. Why should the weak interaction prove an exception?

During the summer of 1956, T. D. Lee and C. N. Yang, two young physicists of Chinese origin working in the United States, began a careful study of all the known experiments involving the weak force. After several weeks' work they reached the conclusion that the conservation of parity in weak interactions had never been firmly established. It was only a principle, not a fact. They published their conclusions in the *Physical Review* of 1 October 1956, proposing several experiments that should make it possible to discover whether the weak force differentiated between left and right. This article went relatively unnoticed, but a physicist at Columbia University, C. S. Wu, took up their challenge. With her team, she set up an experiment on the decay of cobalt 60, which is a highly radioactive isotope of cobalt permanently emitting electrons under the influence of the weak force. The nucleus of cobalt 60 rotates about an axis whose ends may be labelled north and south. The electrons are ejected from the nucleus at its north and south poles. As the orientation of the axis is in no particular direction, the electrons are in fact emitted in all directions. But, if the cobalt is frozen to a temperature close to absolute zero and a strong magnetic field applied to it, nearly all the nuclei may be induced to align their north poles in the same direction. They are said to be *polarised*. These nuclei, which continue of course to eject their electrons, now emit them around two directions only: those in which the north and the south extremities are pointing. If parity is conserved in this type of process, there should be an equal quantity of electrons ejected in both directions.

The experiment showed that things do not happen like this: the electrons are not emitted in the same quantity in both directions. This revolutionary finding came as a bombshell. For the first time in the history of science a means of locating the extremities of an axis had become available which was neither arbitrary nor conventional. It was discovered that, in the weak interactions, nature itself, through its own asymmetry, provided an operational definition of right and left. This result greatly impressed all physicists, and in particular Pauli, who before the Wu experiment had written: 'I do not believe that the Lord is weakly left-handed.' Several other experiments were conducted in the following year, in particular with muons and π mesons, which also decay by weak interaction. In 1958 it became clear that parity is violated in all the phenomena that involve this interaction. The mystery of the tau-theta puzzle was thus explained. There is only one positively charged K meson (designated K^+), and parity is not conserved. For several years the physicists had been knocking on a closed door, and now they discovered that it was not a door, but only the drawing of a door on a wall.

In 1957, Lee and Yang received the Nobel Prize for physics. As Richard Feynman said, they were the first to show that: 'God has made the laws of physics only almost symmetrical, so that we should not be jealous of His perfection.'

There are other symmetries than parity. In a reaction involving particles and anti-particles, it is conceivable that the particles can be replaced by the corresponding anti-particles and vice versa. This permutation between matter and antimatter is called *charge conjugation*, and is designated C. If the transformed reaction has the same probability of being observed as the initial reaction, there is said to be charge conjugation invariance, or invariance under C. It is of course possible to apply the two transformations C and P successively to a given process. If the resultant process is as probable as the initial process, there is an invariance CP. Another operation is time reversal, designated T. It consists of filming a phenomenon and then running the film backwards. If there is nothing to reveal that the film is being projected with a reversal of the direction of time, we have invariance under T. This invariance is very rare for the phenomena of daily life, which involve a large number of particles: apples to not leave the ground and attach themselves to branches (see part two, chapter 7). But many processes involving a small number of particles respect time reversal invariance.

The framework of current theories lays it down that all processes between particles are invariant under the combined operation CPT. In other words, if one runs backwards in a mirror the film of the image of any phenomenon in which there is an exchange of matter and antimatter, the phenomenon observed is as probable as the original one.

After the unforeseen discovery of the violation of parity, the theory of the weak force had to be considerably revised. After much trial and error, the outcome was to assign spatial asymmetry to the structure of neutrinos, which are emitted during decays governed by the weak interaction. The mirror image of a neutrino is no longer a neutrino, but an antineutrino. Thus the new theory abandoned the invariances under P and C separately, but it was agreed to retain the invariance under the combination CP which did seem to be respected.

Things did not rest there, this time because of the neutral K^0 mesons. At the beginning of the 1950s the American physicist Gell-Mann and the Japanese physicist Nishijima had proposed the introduction of a new concept, that of *strangeness*, to deal with the serious identity crisis that seemed to affect the K^0 mesons discovered by Rochester and Butler. Easily produced in high-energy nuclear reactions, and thus by the strong interaction, they decayed 'slowly', with a relatively long average life (of the order of 10^{-10} seconds), characteristic of the weak interaction. Why such schizophrenia? Why did the strong force that presided at their birth not govern their death as well? This strange behaviour was accounted for by attributing to them a new charge, strangeness, analogous to electric charge. But unlike electric charge, which is strictly conserved in all interactions, strangeness is conserved only in the strong and electromagnetic interactions, and not in the weak interaction. Strange particles (not of zero strangeness) decay into non-strange particles (of zero strangeness) under the effect of the weak interaction.

This schema led to curious consequences for the strange neutral mesons, which are the K^0 and its antiparticle the $\bar{K}^0$. They differ only in their strangeness, which is positive for the K^0 and negative for the $\bar{K}^0$. The weak interaction, which is responsible for their decay, is colour-blind to strangeness. It is incapable of distinguishing a K^0 from a $\bar{K}^0$. But is was assumed to respect invariance under CP. This therefore led to the distinction of two classes among the different ways in which the strange K mesons decay. The first class corresponds to those cases where the K^0 decays into states called 'symmetrical under CP', for example into a pair of π mesons. The second class groups more complicated modes of decay, called 'antisymmetrical under CP', which includes decay into three π mesons. It was anticipated that a category of neutral K mesons, designated K_S^0 (S for short), of relatively short lifespan, would decay exclusively according to the modes of the first class, whereas the other K mesons, designated K_L^0 (L for long), with a lifespan 150 times longer, would decay exclusively along the lines of the second class. The experiments made up to 1964 confirmed this analysis, which assumes that the weak interaction respects the symmetry that combines C and P. But in 1964 there was a great surprise. An experiment carried out at Princeton by Christenson, Cronin, Fitch and Turlay showed that, in about one case in a thousand, the long-lived K^0 mesons decay into two charged π mesons. Physicists did not readily resign themselves to this result, which amounted to accepting the violation of CP symmetry, and at first they attempted to interpret the observed phenomenon differently. But all their hypotheses were invalidated by the conclusions of the experiment, and in 1968 came the observation of the decay of the K_L^0 into two neutral π mesons, which drove the point home. They had to yield to the new evidence. CP violation exists weakly for the weak interaction, which has a decidedly fractious temperament. To this day the interpretation of its unorthodox behaviour is still not perfectly clear.

In 1973, two Japanese theoreticians, Kobayashi and Maskawa, showed that, if one postulates the existence of six different quarks (quarks are the constituents of protons, neutrons and all the *hadrons*), a natural explanation of CP violation is obtainable. At the time this was only an academic hypothesis, since only three quarks (*up, down* and *strange*) were known. But two other quarks very soon arrived to extend the list: the *charmed* quark in 1974 and the *beauty* (or bottom) quark in 1977. The last was the sixth or *top* quark, whose discovery was announced in May 1994.

Is it possible to detect a CP violation other than in the decay of K mesons? It does seem so. Experimentally, *beauty* mesons have been discovered, more massive analogues of the strange mesons. To describe them as beautiful is not intended to mean that they are more aesthetic or attractive than other mesons. The beauty they possess is no more than a name for a new kind of charge which, like strangeness or charm, is conserved by the strong and electromagnetic interactions but not by the weak interaction. So we can repeat concerning the B^0 and $\bar{B}^0$ beauty mesons what has already been said about the K^0 and $\bar{K}^0$ mesons. Their decay mode can give rise to the same behavioural oddities. Physicists are so excited by the idea that particles containing beauty can violate CP symmetry that they envisage the construction of *beauty factories* – accelerators specially designed to produce large quantities of beauty mesons (containing a beauty quark).

The subject is not so anodyne as it might seem. For the sacrosanct CPT invariance implies that if CP invariance is violated, then time reversal invariance should also be. This break of symmetry should entail measurable effects, such as the existence of a very small dipolar electrical momentum of the neutron. The electrical neutrality of the neutron would be the resultant of two distributions of charge, one positive and one negative, whose centres would be slightly shifted.

But the most important consequences of CP violation relate to cosmology, the study of the history of the universe. The basic hypothesis of the Big Bang model assumed that matter and antimatter were produced in equal amounts. So where is the universe's antimatter stored?

Twenty years ago it was thought that matter and antimatter had managed to separate very rapidly, without having time to annihilate each other. On this assumption, very far away from us there would exist antigalaxies that we had better not try to visit for fear of exploding on contact. But no experimental argument has emerged to confirm this idea, and in particular no mechanism of rapid separation has been found. Today it is more generally believed that the entire universe consists of matter. But if so, if we are left with nothing but matter, where did the antimatter go?

It is now assumed that matter and antimatter were created simultaneously, with a very slight imbalance in favour of matter. Matter and antimatter then mutually annihilated each other, eventually leaving just a small residue of matter, from which came the galaxies, the stars, ourselves and all the rest. A fortunate imbalance! This formation of matter is called *baryogenesis* (*baryons* are all the particles made up of three quarks, like the proton or neutron). What is its mechanism? We have said that the CP operation brings matter and antimatter into correspondence. An imbalance in the creation of matter starting with primordial energy thus implies a violation of CP invariance. This curious break appeared among the conditions announced by Sakharov in 1967 to explain the mechanism of baryogenesis.

Furthermore, theoreticians have noticed that, if science is not careful, the standard theory of the strong interaction also leads to a violation of CP. No one wants that, because it contradicts all the experimental results. To get round the difficulty, the theoreticians have invented an expedient which consists of introducing a new particle, called the *axion*. As this particle has never been observed, some suppose that it might be found in the *dark matter* of the universe. As its name indicates, dark matter is invisible, but astrophysicists have good reasons for believing that it is hidden in the universe in various forms. By contributing to the total density of the universe, it might determine its long-term destiny.

But this sounds like the start of another fantastic story, perhaps even stranger than that of the mesons called strange.

7

The Paradox of the Arrow of Time, or the Course of History

Time is the meaning of life.
Paul Claudel, The Art of Poetry

'There are many different opinions concerning the essence of time,' said Blaise Pascal. Words that have resisted ... time itself, for time continues to engender unfathomable paradoxes. Physicists, in particular, have been hard put to make their equations disgorge a unique time. Is their time only an illusion? Is it universal? Does it have an arrow, or only the appearance of an arrow?

'Time on your hands', 'in no time at all,' 'in half the time,' 'time out of mind': as witnessed by its presence in so much common parlance, time is built in to our familiar concepts. It also occupies a unique and enormous position in the literature of all ages. How could it be otherwise? It belongs to our most primordial experience of feeling the sensation of a time without which our existence would have neither 'texture' nor 'real life', and to which we would feel helplessly subjected. We are *in* time from the start, and cannot leave it.

Although it is not perceptible by any of our five sense, time is familiar to us. But beware: the concepts most familiar to us are often the most mysterious. In fact we very quickly realise that time is one of the very hardest things to think about, because we cannot *grasp* it, cannot seize the day. We would like to pause and watch the time roll past the way we would a river, without taking part in its flow, but it is tragically impossible. While we are thinking, time sweeps our thoughts along just as it sweeps ourselves. Nothing can stop it passing, and that is time's distinction.

Besides, are we even capable of saying what time is? Is a concept so familiar really graspable by thought? And are we quite sure that we could define time otherwise than by using metaphors of time? Some lovers of verbal gymnastics have seen fit to define it as 'the shrewd means found by nature to make sure that everything doesn't happen at once', but elegant though it is, the maxim does not answer the questions that have plagued so many thinkers. 'What is time and what is its nature? Is it a being? Is it a non-being? Does it imply space? Does it insist on change? And what is its origin?' asked Aristotle in his *Physics*. All of them questions that philosophers have always considered it very hard to answer. So where does that leave us?

Still, let us look a little closer. No sooner do we venture to make sense of time, than we fall among terrible paradoxes. We will look at a few.

– First, there is the problem of the identity of the *now*, which always appeared to us one and the same instant, somehow unchanging. The present is the only thing that has no end and which is always . . . present to us, unlike the past and future, which are only put forward by thought, by way of memory or expectation, but which are never available to us. Aristotle himself (followed by Schopenhauer in particular) underlined this obvious fact that time present is the one true time, but that was so as to go on to contrast this invariance of the present with the mobility of the now: 'Now the instant is, in one sense, the same, and in one sense not; in so far as it varies from one moment to the other, it is different; as to its subject, it is the same' (*Physics*, IV, 219b, 12). So straight away we come up against the primal paradox of time, the one that contrasts the permanence of the *now* with its own dynamic. It doesn't take much shuffling of its contradictory terms to realise that it has an authentic, mind-bending, metaphysical depth.

– Next, there is the problem of the very *reality* of time. As the past no longer is, the future is not yet, and the present itself has already stopped being as soon as it is on the point of starting, how is a *being* of time to be conceived? How could it be that time exists at all, if time consists of inexistences? Let us agree that it would be hazardous to base the reality of time on a reality so vanishingly real as that of the instant: an instant is only a shiver, and the shiver has hardly a grain of ontological weight. And yet, if one had to think that time is nothing, it would also be necessary to agree that there is no change, and that neither growth, nor youth, nor age any longer exists, which would amount to dismissing at a stroke the entirety of our human experience, which is a lot to discard, after all. Conclusion: no more than we can conceive the existence of time, can we conceive its nonexistence! Then what do we say about time, if it is as impertinent to say that it is something as to say that it is nothing? Here we are caught in a seemingly vicious circle, all the more vicious because it is possibly empty. This is the second paradox of time, so intractable that we are struck about as dumb in its presence as Genghis Khan discovering metaphysics.

We will only point out that there are some who believe they can resolve it by assuming, like Aristotle, that 'time is the number of motion, according to the before and the after'; others believe with Saint Augustine that time elapses only in the soul, or else with Kant that it is an a priori form of sensibility, i.e. a kind of empty form. Or ought we rather to see it, as Bergson suggests, purely as an intuition of consciousness? Is there anyone who can answer these questions?

– There is a third time paradox, this one of a semantic sort, which is no less tricky than the first two. That is that the word 'time' says next to nothing about the thing it is supposed to express. This word, which designates the object of an immediate knowledge and experience, fades into mist as soon as one attempts to grasp its content. Plotinus and Saint Augustine were the first to stress this startling contrast between, on the one hand, the distinct spontaneous impression we have of time, and how readily we use the word when we are not puzzling about its meaning, and on the other hand the awkwardness that overtakes us once we look closer. Time is both immediate and ineffable. It can be neither nullified nor grasped. Anything more ambiguous, it is impossible to imagine.

– And that is not all. There is another time paradox, the fourth in this anthology, which is the manifest clash between physical time and subjective time, or if you prefer between *chronos*, clock time, and *tempus*, the time of consciousness, measured in the mind. The first is supposed to be objective and uniform. It does not depend upon us, we are able to measure it, it is displayed on the faces of our watches and clocks. This is the time of our timetables. The second, time experienced, *psychological* time, does not flow evenly. Being so variably fluid, it gives the notion of duration a very relative consistency, so that there are probably no two people who, in a given time, count an equal number of instants. Like elastic, psychological time undergoes twangs, stretches, strains. Paradoxically, the emptier it is the heavier it hangs for us, said Vladimir Jankélévitch, which will serve to differentiate this malleable time – endless for boredom, swift for impatience – from the mechanical time of ordinary objects.

These subjective variations in time perceived, which drags or flies, are so broad that there exist in every life moments of eternity that seem to free it from the tyranny of 'Old Chronos', Father Time. Erwin Schrödinger, a physicist whom we have already mentioned, explained that to experience the timeless, it only needs a circumstance that, without suspending time (which is not a containable flow), comes simply to glue us to the present, and bears us to the centre of time. According to him, it may only take a kiss, which proves that great physicists are not all sheer mind: 'Love a girl with all your heart,' advises Schrödinger, 'and kiss her on the mouth: then time will stop, and space will cease to exist' (notebook of 1919, *On Kantian Philosophy*). Anyone can try.

Whatever the outcome of the experiment, it does seem hard to reconcile these manifold perceptions of time into a single non-reductive synthesis. Then how do we unify *chronos* and *tempus*? Is it even right to wish to make a connection between what is *invariable* and what is *elastic*?

As time is also part of the language of our dialogue with nature, it is not surprising to find it written into all the equations of physics, in the form of the well-known parameter t. This omnipresence of time in the expression of theory raises one immediate question: Is it the mark of a universality of time, or does it reflect a juxtaposition of particular statuses? A second question then arises, with an air of paradox about it (once again!). Is the presence of time in physics not incongruous , in so far as physics tends exactly to deny time by appealing to 'fixed standards' (present in the very idea of a universal law), irrespective of the notion of a dated event?

In order to answer this second question it would be necessary to explore in depth the linkage between the concepts of *history* and *law* (which we shall not go into here). This might make it possible to understand whether physics truly has a mandate to describe the immutable, or whether instead it ought to become the legislation of metamorphosis. Is it the 'formulation of the timeless' or the 'protocol of modifications', to borrow the phrasing of Nicolas Grimaldi in his *Ontology of Time* (1993).

We will attempt no better definitions of the categories we have identified, time 'felt' and time measured. But there is nothing to prevent our comparing them, and that is what we propose to do here, taking as our criterion of comparison that of the reversibility (or irreversibility) of time.

Subjective time appears to us to be irreversible. It has a favoured direction. The past seems to us to be written, fixed, all we can do is remember it, while the future seems uncertain, a priori multiple, still potential. In everyday life, we do not see past and future as symmetrical to one another. We are conscious that the passage of time is inexorable and irreversible. To translate this daily experience, psychological time is said to be *arrowed*. The expression, *the arrow of time*, was coined by the English physicist Arthur Eddington, early in this century.

But from this angle of irreversibility, just what is physical time, or more exactly the different sorts of time envisaged by physics? Do they too have a privileged *direction* of flow? And at what depth are we to seek a correspondence between the respective times of science and existence? Surprising though it seems, it cannot be argued that this problem of the arrow of time has been satisfactorily resolved: developments in modern physics have complicated both the question to be asked and the answers to be given.

We shall begin at the beginning, or thereabouts: with Isaac Newton, who put time 'out of time'.

Newtonian mechanics describes the motion of bodies in space by giving their positions at successive moments. In these calculations of trajectories, time appears as an *external parameter* of dynamics, postulated by Newton as flowing uniformly from the past into the future, which implies that it always flows in the same direction, and is therefore arrowed. But, curiously, this time is reversible because past and future can be explored using the same mathematical methods. With every movement from past into future, mechanics associates a symmetrical movement from future into past: it is just as easy to calculate past eclipses as future eclipses, and, on paper, the planets could just as well spin backwards. The reversal of time, in other words exchanging past and future, has no effect on the equations. Hence Newtonian time has no *arrow*; it neither creates nor destroys. It merely measures the pace and calibrates trajectories. Newton invented a scrupulously neutral time. All in all, Newtonian mechanics makes no distinction between past and future, which it reduces to a single present moment, as described by Newton's successor, the Marquis de Laplace, in the famous passage quoted in part one, chapter 6.

This absence of a Newtonian arrow of time is a further paradox, or at any rate was experienced as such by a constellation of nineteenth-century scholars such as Boltzmann, Gibbs, Zermelo, Loschmidt and Poincaré, and more recently by Prigogine (Nobel prizewinner for Chemistry in 1977). For there genuinely are events that can only act in one direction, in other words in phase with an arrowed one-way, time. Furthermore, the great majority of the events we observe are irreversible (as Marcel Proust took his time to express). No one has ever seen a cup of coffee reheat itself spontaneously, or a living thing grow younger, and it is well known that in general 'history does not give second helpings'. But what kind of stuff is this 'arrow'? What temporalises physics?

Focusing on just such irreversible processes, the nineteenth-century physicists Sadi Carnot and Rudolf Clausius discovered the second law of themodynamics. This macroscopic law first postulates the existence, for any physical system, of a quantity – which Clausius called *entropy* – fixed by the energy state of the system, which broadly represents the degree of disorder or randomness in the system (it is a state quantity like volume, pressure or temperature). The second law states that the amount of entropy contained in a closed system can only increase in any physical process. (Woody Allen has Judy Davis define it more graphically in *Husbands and Wives*: 'Sooner or later everything turns to shit.') As entropy can increase only *in the course of time*, then time is arrowed, it flows in a preferential direction, and therefore the past and the future are differentiated. The growth in the entropy of a closed system simply expresses the average tendency displayed by that system to evolve at the molecular level towards more and more probable states, which means states of increasing disorder. The second law therefore seems to tally closely with our sense of time having a regular direction. On first sight, at least . . .

In reality, we have to look more closely. Among the equations of physics, there are some that are fundamental, in the sense that they account for the basic behaviours of matter and, *in principle*, explain everything. They are known as *microscopic*, because they deal essentially with the elementary bricks – atoms or other particles – with which the whole universe is supposed to be built. The key point here is that all the microscopic equations of physics are reversible: when the time variable is run in a particular direction, say towards the future, the equations describe a certain motion for particles; when it is run the other way, the motion calculated is the same as before, but described in reverse. Neither motion can be said to be more or less physical than the other (the reverse of two billiard balls colliding is also two billiard balls colliding), so there is no way to determine which of the two corresponds to the true flow of time. In other words, if you film a microscopic event and then run the film backwards, which is the equivalent of reversing the chronology of the event, no spectator can tell that you have intervened. That is why microscopic equations are said to be reversible. In sum, they describe particles that do not age. But beside these microscopic equations, there is another less fundamental kind that, as it were, sum up a more wholesale behaviour of matter. These equations are called *macroscopic* because they describe phenomena that happen on a scale close to our own. In principle, the macroscopic equations should flow straight out of the microscopic equations, since net behaviour is no more than the combination of a large number of elementary events.

Meaning to go deeper into this question, Ludwig Boltzmann (1844–1906) tried to find a link between (Newtonian) mechanics and the above-mentioned second law of thermodynamics. On the basis of the reversible microscopic equations, could he derive, by 'aggregation', an irreversible macroscopic equation? As rigorous integration of the behaviour of a very large number of particles is impossible. Boltzmann resorted to the laws of statistics, and this enabled him to study the movement towards thermal equilibrium of the distribution of the velocities of molecules in a gas. He stated that a quantity H representing this distribution has a remarkable property: it can only decrease, to approach a minimum when it reaches equilibrium. It is therefore the microscopic analogue (apart from the sign) of entropy. Thus the statistical aggregation of the equations of particle dynamics does lead to an irreversible macroscopic equation. The longed-for arrow seems to soar – almost miraculously – out of a thicket of calculations.

Does that settle the problem of temporal assymetry?

Certainly not, because this calculation interprets irreversibility as being only a statistical property of macroscopic systems, i.e. those containing a very large number of degrees of freedom. By this interpretation, irreversibility is only a statistical illusion, while the 'real', microscopic, reality remains reversible: there is no arrow of time, only the *appearance* of an arrow. From there to asserting that time itself is only an illusion is one short step that some theorists, and not the least among them, have gladly taken. Einstein himself wrote in his private correspondence (letter written after the death of his friend Michele Besso to Besso's family) that 'for us physicists, the distinction between past, present and future is a mere illusion, even if a stubborn one.' Though his outlook on the question was not always so radical, the fact remains that Einstein had high hopes of eliminating the notion of irreversibility by restoring physics to pure geometry, in other words to a form 'without history'.

By contrast, other physicists doggedly refuse to forsake as illusion the time that they consider as the paramount experience of our life. They claim that a 'true' time really does elapse, which physics has failed to see and integrate. This is the view of Ilya Prigogine, who set out in search of the lost time of physics. He takes the example of the molecules forming the contents of a glass of water. Does the glass of water age? Yes, replies Prigogine. The molecules present in the glass collide among themselves, which sets up correlations between them, just as the meeting of two people leaves a memory. Reverse the velocities of the molecules, and they will re-collide and their correlations will propagate. According to Prigogine, this arrow of correlations must correspond to a true arrow of time.

So instead of saying: 'There is no arrow of time, but the macroscopic level creates the illusion that there is', we can follow Prigogine (but without necessarily endorsing his arguments), turn the argument on its head, and say: 'There is an arrow of time, but the microscopic level creates the illusion that there is not.' In which case irreversibility ceases to be a matter of outlook, and becomes inherent in nature. The second law of thermodynamics becomes the first law of physics as a whole; the concept of *becoming* (the growth of entropy) transcends that of *being* (particles and other physical objects); the idea of *evolution* prevails over that of *existence*. By virtue of this reversal, which resembles a Copernican mini-revolution, the concept of the 'historicity' of the Universe would come to the forefront of physics. The notion of the *age* of systems would also acquire meaning, at least for systems not in equilibrium. But it remains to be seen how the arrow of time would manage to puncture the harmonious edifice of classical mechanics, so notoriously indifferent to the message of irreversibility attached to it.

Especially because classical mechanics is not the only runner. Modern physics is also special relativity, general relativity, quantum mechanics, field theory, cosmology, and so on. Einstein, for example, does not postulate time as an a priori. Very pragmatically, he focuses on the durations measured between two events by observers situated in different frames of reference, and shows that these measurements do not yield the same results. This is the great lesson of special relativity (1905), which put an end to the reigning confusion about the ether (the Maxwellian version of Newton's absolute space). It obliged Einstein to introduce the concept of 'space-time' to replace the hitherto separate concepts of space and time: shifting the frame of reference in space-time transforms time partly into space, and space into time. In fact, in order to 'move' from one system of space-time coordinates to another in uniform translation in relation to the first (i.e. neither accelerated nor slowed), we use what is known as a 'Lorentz transformation', which has the characteristic of 'blending' spatial and temporal coordinates with no real distinction. So there is no question of letting time take care of itself, and in any case Einstein is utterly opposed to it. Time and space become relative, which contravenes the mental habits of millennia. As a philosophical consequence, time loses its Newtonian ideality, stops being external to space, and starts to depend on dynamics. As a practical consequence, clocks in rapid motion tick more slowly. This slower tempo, which measures the 'elasticity' of relativistic time, is commonly observed in unstable elementary particles such as muons (see part two, chapter 2).

As long as we took for granted a universal time, it was possible to say that the past no longer exists, the future is still to come, and therefore only the present exists (even if it lasts for just a moment). Relativity undoes this proposition: events that are in the future for one observer are in the past for another and in the present for a third. What is more, there are nowadays as many fundamental chronometers as there are objects in uniform motion. And these clocks cannot be synchronised. If their faces or displays are adjusted at a given moment, the hours indicated no longer coincide a few moments later. Time no longer has a standard, and the concept of simultaneity is losing its meaning. In such a tangled context, what becomes of the proper status of time?

Let us not even mention, or hardly mention, Einstein's General Relativity, which seems today to be the right theory of gravitation. It features a curved space-time within which space, time and *also* matter fuse in very complex collusion. Einstein's equations propose in fact that the density of mass and energy conditions the very structure of space-time, and that it is this structure (known as the 'metric' of space-time) that correspondingly determines the dynamics and traject-ory of the objects contained in the universe. In such a context, mass directly influences the speed of the pas-sage of time. In cosmology we also speak of an arrowed *cosmological* time, useful for projecting the evolution of our Universe from the first (but not quite first) moments of the Big Bang. But this time cannot be posited as an outright absolute: as it depends upon the theoretical model chosen, it is difficult to consider it as the framework where our perceptions simply drop in and settle down.

Lastly there remains the – very delicate – case of quantum mechanics. For simplicity's sake, we shall single out the Schrödinger equation (not a relativistic constant), which enables us to calculate the behaviour of the wave function associated with every particle. This equation is perfectly reversible, and equally per-fectly deterministic. The time it deals with is therefore a priori Newtonian time. Actually, things are not so straightforward, for quantum physics has a twofold structure: *as well as* its equations, it also requires a theory of measurement, in other words a means of relating its formalism to practical results. In fact, the Schrödinger equation is valid for as long as the system goes unmeasured. If a measurement is made, only one of the outcomes possible a priori actually happens, and it is selected strictly by chance. Measurement randomly 'actualises' just one of the system's potentialities (this is 'quantum indeterminacy'). Everything works, in fact, as if the act of measuring presents the system with a choice of several possible futures, and at the moment of measurement chance volunteers to select one. We say that there has been a 'collapse of the wave packet'. Does this process, or does it not, introduce a temporal irreversibility? If it does, it would be a very odd arrow that flew here: the performance of measurements on systems would be intervening implicitly in the creation of irreversibility!

One last thing: particle physicists wishing to test matter–antimatter symmetry discovered to their great surprise in 1964 that the weak nuclear interaction (responsible for the phenomena of beta radioactivity) conflicts very slightly with past–future temporal symmetry (see previous chapter). The question remains to see whether or not to associate this breach of symmetry with an arrow of time. Today, most of the physicists who are investigating the arrow of time from a synthetic perspective think that it may be connected with gravitation. In fact it may well be that there is a unifying link between violation of matter–antimatter symmetry, gravitation, and the arrow of time.

For the moment, however, what is clear is that things are unclear. Physics has at least four times. Each of its conceptual systems gives time a distinct and original status. Hence time itself wears the face of the sphinx, and its nature remains the sphinx's riddle. Would this mystery be dispelled if physicists succeeded in unifying the four fundamental interactions that they have iden-

tified? After all, all of these times that we have mentioned may contain a hard (but cryptic) kernel of
common properties. For physics to decode the 'signs
of the times' would unquestionably throw new light
on many fundamental problems – for example on the
interpretation of quantum physics.

As for irreversibility, is it mirage or reality? The
proper answer to this question will no doubt reveal
the bonding between external and internal factors, the
nature of the contrast between objective and subjective
time. But what has already emerged quite distinctly is
that it will take more time to clarify time, even if only
to explain that it does not exist. More time, but also
more space: relativity requires it . . .

Conclusion

Your brain, doctor, is a culture medium for
question-marks!
Paul Valéry

We could continue to itemise the lengthy list of the
paradoxes that have left their mark on physics, but
apart from the fact that we would then run the risk of
boring the reader, that would be an infinite task. In
order to be comprehensive, we should have to write
nothing less than the entire history of physics, which
is beyond our power and competence, and incompatible with our present agenda. Nevertheless, these seven
examples suffice to show the richness and diversity of
the concept of paradox. Brief though it has been, this
account has allowed us to survey what they involve
and to look deeper into their true nature.

Paradoxically, paradox itself is also the victim of
prejudice: it too is often reduced to the idea of contradiction, failure, conflict. And as science tends to be
identified as a solid receptacle for certitudes, some
people leap too quickly to the conclusion that the
existence of a paradox in physics is a sign that the
theory is defective, or has not yet fully shaken down,
and that it is by eliminating the paradox that it will
become complete.

Viewed in that light, paradox is basically a kind of monster that has to be driven away.

But that is not the way things work. Paradox is a healthy animal. It forms an integral part of scientific progress, to which it gives rhythm and rigour. It debunks false consensuses, opens perspectives, initiates new lines of approach, gives ideas the chance to come forward and be tested, and suggests viable conceptual grafts. It is a challenge to the intelligence, which spurs the imagination and intensifies the urge to understand.

Paradox is vital for science, the place where it dwells. It must no longer be seen as an unfortunate accident that more attention, care or luck would enable us routinely to avoid. Physics, like every activity of thought, feeds on transcending its own tensions. It is from the springboard of paradoxes, and the problems they present, that it has generally progressed.

Iconoclastic and provocative, paradoxes prevent science from relapsing into a platform for issuing *ex cathedra* statements. Without paradoxes there would be only a closed, formal, desiccated, repetitious science. They are to science what a brilliant twist is to a good play – a means to galvanise interest and revivify the plot.

Even after they are cleared up, paradoxes retain all their value and freshness. They remain phenomenal teaching tools that ought to be used more often in education. It is one thing to master, for instance, the formalism of the special theory of relativity, another to understand Langevin's paradox. Getting the hang of a paradox fires the whole of the intelligence in a way much more intense than the processes of calculation, or operational routines. When one has made the effort to understand a paradox, it illuminates the whole of the theory that gave rise to it. Paradox is the tool of clarity, a theoretical beacon.

Lastly, paradox is a place of encounter and emergence, as well as a hymn to polyvalence, multiplicity. In order to grasp and thoroughly analyse a paradox, we have to ask a series of questions that take us far beyond the technical or formal instance that conveys it. When does it date from and what does it consist of? What reactions has it provoked, who discovered it, and how was it surmounted? What new ideas did it engender? The complete anatomical study of a paradox says much about the genesis of ideas and the mechanisms of human thought. Paradox is an inter-disciplinary crossroads, a creative confluence. It summons and assembles all the conquests of the intellect – science, history, epistemology, sociology, philosophy, humour – and in the last analysis, it perfects them.

From oxymoron to the riddles of quantum theory, paradox is everywhere, plural and polymorphous, banal or profound.

Wherever it appears, it causes the desert to blossom.

Bibliography

Caussat, Pierre (1992). *L'Événement*. Desclée de Brouwer.

Chalmers, Alan F. (1978). *What is this Thing Called Science?*. Open University Press.

d'Espagnat, Bernard (1976). *Conceptual Foundations of Quantum Mechanics*. W. A. Benjamin.

—— (1990). *Penser la science ou les Enjeux du savoir*. Dunod.

—— (1994). *Le Réel voilé*. Fayard.

Falleta, Nicholas (1986). *Paradoxicon: A Collection of Contradictory Challenges*. Turnstone Press.

Gardner, Martin (1967). *the Ambidextrous Universe*. Penguin Books.

Heisenberg, Werner (1971). *Physics and Beyond*. Allen & Unwin.

Hoffman, Banesh (1976). *Albert Einstein: Creator and Rebel*. Paladin.

Klein, Étienne (1994). *Sous l'atome les particules*. Collection 'Dominos', Flammarion.

—— (1995). *Le temps*. Collection 'Dominos', Flammarion.

Kuhn, Thomas (1976). *The Structure of Scientific Revolutions*. University of Chicago Press.

Ortoli, Sven, and Pharabod, Jean-Pierre (1985). *Le Cantique des quantiques*. La Découverte, Le Livre de Poche no. 4066.

Schrödinger, Erwin. *What is Life* and *Mind and Matter*. Cambridge University Press.

Stengers, Isabelle, and Schlanger, Judith (1989). *Les Concepts scientifiques*. La Découverte.

Verdan, André (1991). *Karl Popper ou la Connaissance sans certitude*. Presses Polytechniques et Universitaires Romandes.

Index